Damaged Ship V

London, United Kingdom
11 March 2020

ISBN: 978-1-7138-0865-7

Printed from e-media with permission by:

Curran Associates, Inc.
57 Morehouse Lane
Red Hook, NY 12571

Some format issues inherent in the e-media version may also appear in this print version.

The Royal Institution of Naval Architects
8-9 Northumberland Street
London, WC2N 5DA
United Kingdom

Phone: 020 7235 4622
Fax: 020 7259 5912

publications@rina.org.uk

Additional copies of this publication are available from:

Curran Associates, Inc.
57 Morehouse Lane
Red Hook, NY 12571 USA
Phone: 845-758-0400
Fax: 845-758-2633
Email: curran@proceedings.com
Web: www.proceedings.com

International Conference

Damaged Ship V

11th March 2020, London, UK

PAPERS

THE ROYAL INSTITUTION OF NAVAL ARCHITECTS
8-9 Northumberland Street, London, WC2N 5DA, Telephone: +44 (0)20 7235 4622

The Royal Institution of
Naval Architects

INTERNATIONAL CONFERENCE

Damaged Ship V

11th March 2020, London, UK

CONTENTS

STABILITY CONSIDERATIONS FOR OFFSHORE TRANSPORTATION VESSELS

A.P. Crowle, McDermott, UK

SUMMARY

Offshore transportation vessels are used for the ocean delivery of structures for oil and gas production and also for renewable energy facilities. The paper will consider the requirements of Marine Warranty Surveyor stability guidelines for temporary conditions. Both self propelled Heavy Transport Vessels and towed Cargo barges are included.

The paper will discuss intact and damage stability naval architecture calculations for the temporary conditions of loadout, ocean transport and installation. The implication of the temporary conditions and in particular transport vessel ballasting, with regards to overall and local strength for the cargo and transport vessel will be considered.

The paper will also provide methods and analysis results for improving intact and damage stability by using sponsons, stability boxes or stability pontoons. The options available for ballasting will be discussed. Previous incidents will be reviewed and lessons learnt described.

The paper will consider initial GM, range of stability for the intact condition and in addition for the damage cases angles of heel and trim and changes to draft after damage. Damage will be considered due to collision, grounding, progressive flooding, fire, explosion and the effects of incorrect-ballasting during marine operations.

NOMENCLATURE

[Symbol]	[Definition] [(unit)]
COG	Centre of Gravity
CUM	Cubic metres
Deg	Degree
DP	Dynamic positioning
FSRU	Floating, Storage, Regas Unit
GM	Metacentre above COG
HAT	Highest astronomical tide
HAZID	Hazard indentification
HTV	Heavy Transport Vessel
LAT	Lowest astronomical tide
LOA	Length overall
M	Metre
SPMT	Self propelled modular transporter
SSCV	Semi submersible crane vessel
T	Tonnes

1.0 INTRODUCTION

The temporary phases of an offshore structure transport requires specialised vessel. The transport vessels may be powered, Heavy Transport Vessel (HTV) or require towing, a cargo barge. Intact stability, both GM and GZ curve, and damage stability need to be considered. This paper outlines the temporary phase definition, marine warranty requirements, typical transport vessels, incidents and stability enhancement. This paper covers project cargo used for onshore modularisation projects, renewable (in particular transformer platforms for offshore wind farms, offshore modules and floating cargo.

2.0 TEMPORARY PHASES

2.1 Loadout

Loadout is the movement of the Cargo from a construction yard onto a transport vessel.

Input requirements are for the loadout quay are:

- Loadout method Skidding, SPMT or Lifting
- Quay height above lowest astronomical tide (LAT)
- Water depth at the loadout quay relative to LAT
- Tidal range, HAT to LAT
- Mooring bollards on the quay
- Mooring winches on the quay
- Lightship weight and COG

Input requirements are for the ocean transport vessel are:

- Cargo weight and Centre of gravity (COG)
- Upper grillage weight and COG
- Ballast tank arrangement
- Ballasting method on the transport vessel which can either be using vessel pumps, external ballast system or a combination
- Mooring bollards on the transport vessel
- Lightship weight and COG

Criteria

Loadout stability criteria are:

- Transport vessel stability has to be adequate throughout the loadout operation.
- Loadout minimum GM is to be 0.15m but a more realistic minimum GM of 1.0 m is expected at all stages. The accuracy requirements to ballasting will tend to increase

with decreasing GM. The GM calculation must include free surface effects

- A loadout with a small metacentric height, where an offset centre of gravity check if a small heel or trim occurs as the cargo transfer is completed
- A loadout where there is a significant friction force between the transport vessel and the quay wall, contributed to by the reaction from the pull on system and the moorings. The friction may cause "hang-up" by resisting the heel or trim, until the pull-on reaction is released, or the friction force is overcome, whereupon a sudden change of heel or trim may result.
- Cases where a change of wind velocity may cause a significant change of heel or trim during the operation.

Assumptions

During a loadout other marine traffic is kept away, thus minimizing the possibility of collision and waves over topping the deck of the transport vessel when it is at low freeboard.

During loadout the under keel clearance is at least a minimum of 1.0m, thus minimizing the possibility of grounding.

Freeboard

The effective freeboard during the load-out shall always be greater than:

- 0.5 m plus 50% of the maximum wave height expected during the load-out operation if procedures are in place to monitor the freeboard at all four quarters of the transport vessel
- 0.8 m plus 50% of the maximum wave height expected during the load-out operation if no freeboard monitoring is implemented

Assume extreme significant wave height, Hs during loadout is 0.5m, i.e. maximum wave height

= 1.86*0.5 = 0.93m

Thus minimum freeboard at all corners

= 0.5 +0.93*0.5 = 0.965m

The effective freeboard allows for the maximum possible tide level during the load-out period. The bunding of openings in the transport vessel deck is required for low freeboards.

Ife the load line is exceeded, then the minimum freeboard used during the loadout operation is confirmed as acceptable by the transport vessel's class society and is to be sufficient to maintain the transport vessel's water-plane area. A semi-submersible HTV or barge will already have dispensation to ballast above the loadline in weather limited marine operations. The loadout is carried out in wind speeds less than 10m/s and significant wave heights less than Hs=0.5m

Damage

Normally there is no requirement to document damage stability during loadout. However, consideration of /how incorrect operation of the ballast system may influence stability is to be checked.

2.2 Ocean Transport Vessel

Input requirements during ocean transport and coming alongside the offshore installation vessel:

- Cargo weight, COG and radii of gyration
- Upper grillage weight, COG and radii of gyration
- Lower grillage weight, COG and radii of gyration
- Seafastenings weight, COG and radii of gyration
- Ballast tank arrangement
- Ballasting methods
- Minimum water depth on the ocean route
- Air draft restriction, bridges and power cables
- Width restriction, locks and canals
- Lightship weight, COG and radii of gyration

Ballasting

The following constraints are considered:

- Ensure sufficient intact and damage stability
- Trim angle is less than 1%
- Heel angle is zero
- Minimize motions and accelerations on the cargo. Thus the natural roll period should be away from the expected wave periods
- Longitudinal shear forces and bending moments are within allowable
- Local strength is acceptable. Lower grillage will probably be required to spread the loads from the cargo into the deck of the transport vessel
- Minimize powering requirements for HTV or the towing tug

The natural roll periods is given by

Tr= C*beam*beam/GM

Where GM includes free surface effects and typical values are k is between 0.35 and 0.6.

Intact Stability

The transverse metacentric height (GM) must be 0.15m and preferably 1.0m, at zero angle of heel. Correction to values of GM to allow for free surface effects is to be be included in this calculation.

The application of free surface corrections to reduce metacentric height (GM) and hence to increase natural roll period will not generally be accepted. The effect of any reduction in GM due to free surface effect is to, be considered in intact and damage stability calculations.

Classification Societies, Marine Warranty Surveyors and Statutory Authorities each have their own set of requirements to be met by the stability curve.

The intact stability, or intact range of stability, is the range between 0° heel or trim and the angle at which the righting arm (GZ) becomes negative

Figure 1 Righting Arm

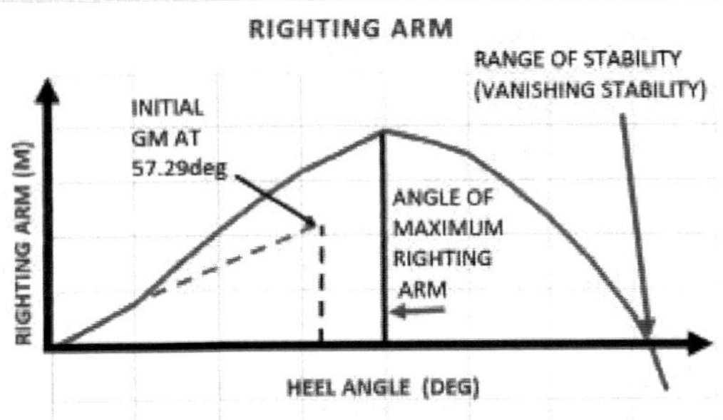

The acceptable range of intact stability for various vessel and barges are:

VESSEL	LENGTH OVERALL	MOULDED BREADTH	INTACT RANGE OF STABILITY
Large vessels	Greater than 76m	Greater than 23m	36 deg
Large barges	Greater than 76m	Greater than 23m	36 deg
Small barges	Less than 76m	Or less than 23m	40 deg
Small vessels	Less than 76m	Or less than 23m	44 deg

The acceptability of small barges with a range of 30 to 40 degrees will be dependent on motion response predictions. Similarly the acceptability of large barges with a range of 30 to 36 degrees will be dependent on motion response predictions

In the event of the range of static stability being greater than 30 degrees and less than 40 degrees, it is to be demonstrated that the maximum predicted roll angle is less than the angle at which the maximum righting lever occurs.

ISO Marine Operations, Ref 1, The stability range should not be less than

$$20 + 0.8*\theta$$

where θ is the sum of the static wind heeling angle and the maximum roll angle in degrees. Where cargo overhangs are immersed as a result of heeling due to a 15 m/s beam wind in still water conditions, it is demonstrated that transport vessel controll is not seriously impaired, and that no structural damage to the cargo can occur.

For marine operations of very short duration in sheltered waters (for instance harbour moves and out-of-dock operations) that are covered by a reliable weather forecast, an exemption from the intact stability requirements may be considered. However, the stability range must never be less than 15°.

Alternatively for barges, Ref 2, if maximum amplitudes of motion for a specific towage or voyage can be derived from model tests or motion response calculations, the intact range of stability shall be not less than:

$$15 + (15/GM) + \theta$$

where GM is in metres and θ = the maximum amplitude of roll or pitch caused by the design seastate plus the static wind heel or trim caused by the design wind, in degrees. A range of static stability less than 30 degrees is not normally be accepted. Typical stability requirements are:

INTACT STABILITY	MINIMUM	UNIT
Residual Area from 0 deg. to Max. Righting Arm	➢ 0.0630	m-Rad
Residual Area from 30 deg. to abs 40 or Flooding angle	➢ 0.0300	m-Rad
Residual Area from 30 deg. to abs 40 or Righting arm zero	➢ 0.0300	m-Rad
Righting arm at 30 deg.	➢ 0.200	m
Angle from 0 deg. to Max Righting arm, for vessels > 150m length	➢ 15.000	deg
Angle from 0 deg. to Max Righting arm, for vessels < 100m length (interpolate lengths 100 and 150m)	➢ 20.000	deg
GM upright shall be	➢ 0.150	m
GM upright should be	➢ 1.000	m
Righting Arm ratio to wind moment	➢ 1.400	

Intact Dynamic Stability

The areas under the righting moment curve and the wind heeling moment (or wind moment) curve is calculated up to an angle of heel which is the least of:

- The angle corresponding to the second intercept of the two curves.
- The angle of down flooding.

Figure 2 Intact area under righting moment curve

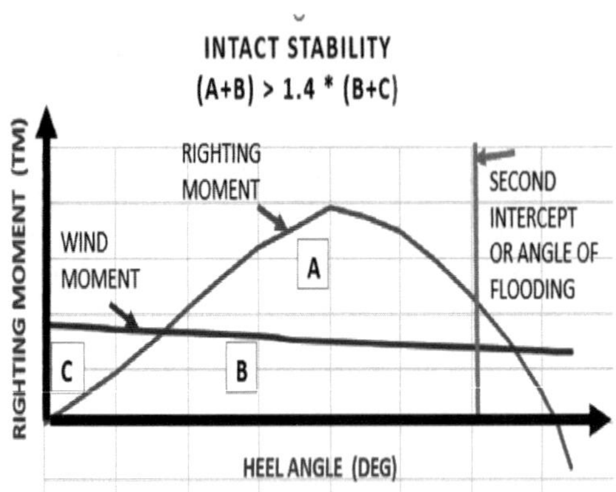

The wind velocity used to compute the wind heeling moment curve should be the 1minute sustained wind for the operation with return periods. The duration is the tow time plus worst case waiting on weather at the offshore installation site. The default return period is 10 years return period.

DURATION OF VOYAGE	DESIGN WAVE AND WIND CRITERIA
Up to 3 days	Specific monthly weather criteria
3 days to 1 week	1 year return period and seasonal
1 week to 1 month	10 year return period and seasonal
1 month	10 year return period, seasonal
Seasonal = summer, autumn, winter or spring	

Assume the wind velocity used to compute the overturning moment curve should be 36 m/s (70 knots), if no data is available.

Damage stability

As a minimum, the transport vessel should have sufficient stability and reserve buoyancy to remain afloat at a waterline below any opening where progressive flooding may occur with any one compartment adjacent to the sea flooded.

Damage to any compartment above the intact waterline that can lead to loss of stability is considered when assessing damage stability.

The loss of water from a full compartment is considered if it gives a more severe result than the flooding of an empty compartment.

The areas under the righting moment curve and the wind heeling moment curve is calculated from equilibrium up to an angle of heel which is the least of:

- The angle corresponding to the second intercept of the two curves.
- The angle of down flooding.

The wind velocity used to compute the overturning moment curve is 25.8 m/s (50 knots).

Where it is impracticable to comply with damage stability recommendations, a hazard risk assessment (HAZID) is carried out, and appropriate mitigating measures taken.

If buoyancy of the cargo has been included to meet intact stability requirements, then loss of cargo buoyancy or flooding of cargo compartments, shall be considered as a damage case, as appropriate.

Typical damage stability criteria are:

- Area (A+B) > 1.4* (B+C)
- Positive freeboard

The damage stability requirements depend on the:

- Transport vessel
- Voyage route
- Summer or winter voyage

Figure 3 Damage stability dynamic stability

There is a negative GM at the zero heel angle

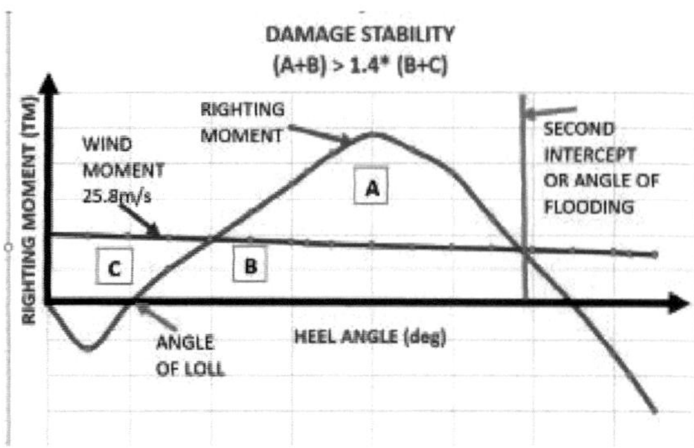

2.3 Allowable VCG curve

The allowable VCG curve for the transport vessel is used by non naval architects to understand the stability restrictions. It combines the requirements of intact and damage stability.

Figure 4 Typical allowable VCG curve

2.4 Ocean Transport Vessel for floatover

Input requirements are during ocean transport and coming alongside the offshore installation vessel:

- Cargo weight, COG and radii of gyration
- Upper grillage weight, COG and radii of gyration
- Lower grillage weight, COG and radii of gyration
- Seafastenings weight, COG and radii of gyration
- Ballasting methods
- Minimum water depth at the installation
- Under keel clearance

Stability requirements during floatover are:

Criteria

Transport vessel stability has to be adequate throughout the floatover installation operation, namely:

- Stability checks are carried out for the full range of probable GM values, module weight and centre of gravity predicted during installation this must include the effects of deballasting the barge and jacking the module where applicable.
- Any installation with a small metacentric height, where an offset centre of gravity, of the Cargo, may induce a heel or trim during the ballasting / weight transfer
- Cases where a change of wind velocity or wave direction may cause a significant change of heel and trim during the installation.
- During float-over installation it is necessary to maximise float-over clearance by maximising the transport vessel draft within stability limitations. For this case only intact stability need be considered with a positive GM of 1.0m
- Down flooding points should be above wave crest heights after allowing for run-up.

Assumptions

It is assumed that the minimum required GM is 1.0m at all stages of the floatover, after the effects of free surface have been considered.

It is assumed that there is always 1.0m of under keel clearance during all stages of floatover

The stability of the vessel during the floatover, especially in the case of a ballasting malfunction is to be considered.

Draft And Freeboard

The maximum draft of the installation / removal barge during float-over shall not exceed the maximum loadline draft, without a class exemption (but this is not normally needed for semi-submersible heavy transport barges or vessels). For operations involving semi-submersible barges or heavy transport vessels with watertight main decks, wave crests may be allowed to over-top the vessel deck provided that all hatches and down flooding points are suitably protected and that raised walkways are added to all areas affected by water on deck where personnel movement is required.

The minimum freeboard is defined as the minimum distance from the waterline to the watertight deck level after accounting for static trim and heel. The minimum freeboard shall be sufficient to maintain the vessel's water-plane area and to ensure sufficient stability range at all stages of the operation, the minimum freeboard isbe at least 1.0m. A lower minimum freeboard may be acceptable if adequate precautions and procedures are in place to ensure that the stability of the vessel is maintained. The minimum freeboard used during the operation shall be confirmed with the barge owner.

Permeability

Examples of permeability to be used in damage condition:

- Living Space 95%
- Machinery Space 85%
- Ballast tank 99%

3.0 TYPICAL TRANSPORT VESSELS

3.1 Cargo barges

Cargo barges have no propulsion and so need ocean tugs to tow them offshore.

Parallel sided barges are used to transport deck cargo. Loadout is usually over the stern. See figures5, 16,17 and 18, which show parallel sided barges.

Figure 5 Parallel sided transport vessel

Parallel sided barge are fitted with skid rails for launching jackets. Loadout is then over the stern.

T shape barges are designed to transport modules for floatovers. The stern is much wider than the bow. These barges are fitted with skid rails for launching jackets. Loadout is only over the stern. See figures 6 and 19 which T shaped barges.

Figure 6 T shape barge

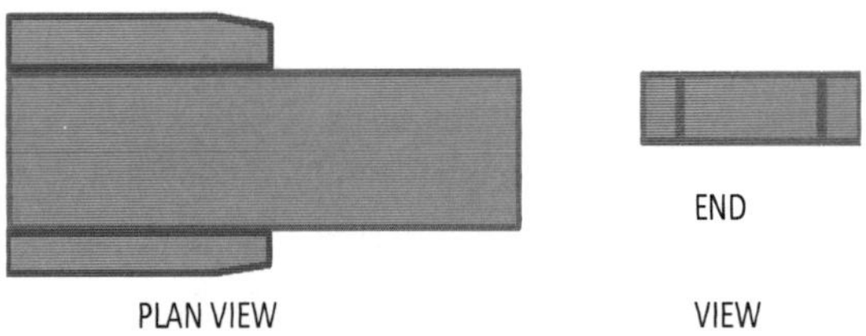

Submersible barges are designed for loading or off loading floating cargo such as semi submersibles or jack-ups. They can also be used to transport dry deck cargo. Loadout can be over the stern or side. Floating discharge is in shallow water, as the stern will rest on the seabed.

Figure 7 Submersible barge, no accommodation

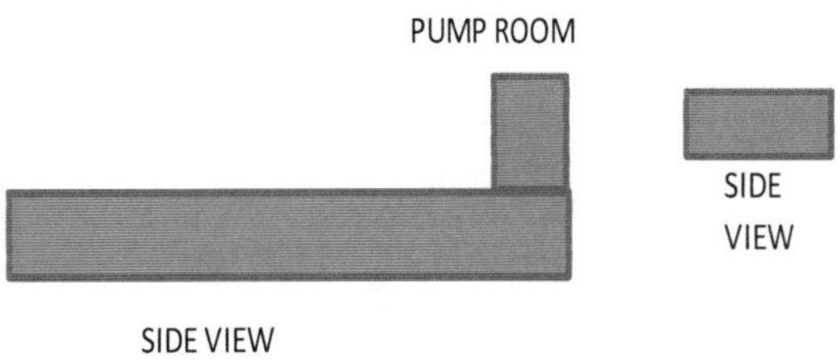

3.2 Heavy transport vessels

Heavy transport vessels have self propulsion. Some may have DP 2 systems.

Geared HTVs are fitted with high capacity cranes. Loadout is by crane. They may be fitted-out with stability pontoons, which are deployed on the crane side of the Geared HTV.

Module carrier HTV, figure 8, cannot be submerged and are suitable only for dry deck cargo. Loadout is generally over the stern, by SPMT or skidding.

Figure 8 HTV module carrier

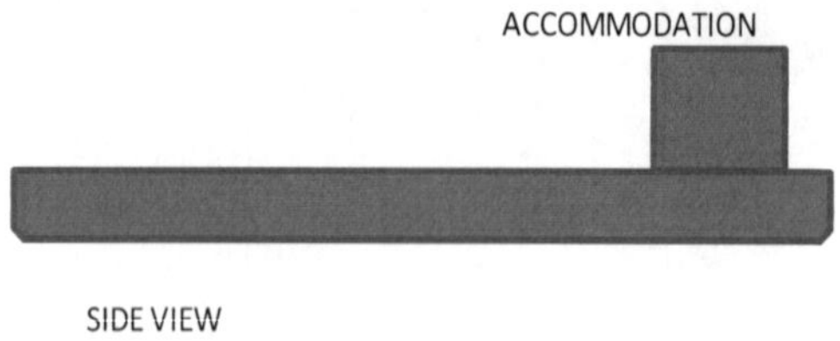

Double ended HTV, figure 9, are designed primarily for loading or off loading of floating cargo. Where a dry cargo is transported loadout is over the side only.

Figure 9 HTV Double ended

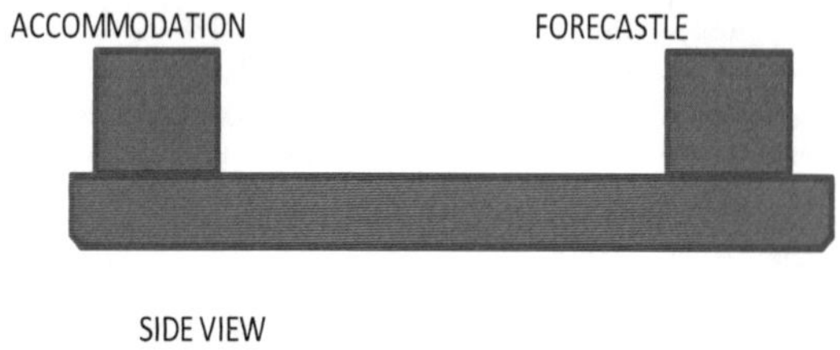

Single ended HTV, figure 10, can be used for loading or off-loading of floating cargo when fitted with stern buoyancy tanks. If the buoyancy tanks are removed from the sternthen the HTV can be used for transporting dry cargo, The cargo can be loaded over the side in moderate tidal ranges, of say less than 2.5m. Stern loadout is possible assuming the stern is of sufficient strength.

Figure 10 HTV with removable stern buoyancy tanks

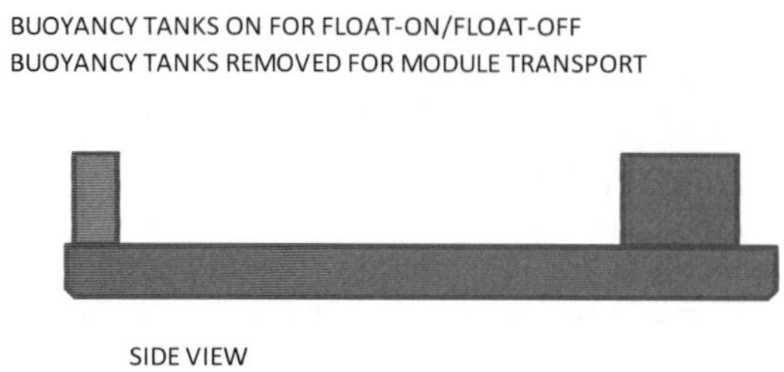

4.0 STABILITY ENHANCEMENT

4.1 Sponsons

Sponsons tanks provide buoyancy and stability enhancement.

Sponson Tanks, figure 11, are projections from the sides of a vessel and they acts for protection, stability, or mounting of equipment such as armaments or lifeboats, etc. They extend a hull dimension at or below the waterline and serve to increase flotation or add lift when underway.

Figure 11 Sponsons added

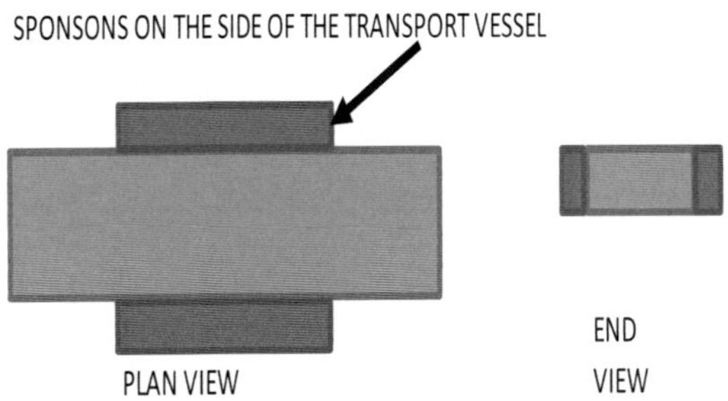

Disadvantages:

- Increased transport vessel weight
- Dry docking to fit the sponsons
- Dry docking to remove the sponsons
- Increased drag, slower steaming speed
- Scrapped after use

4.2 Stability boxes

Stability boxes, provide stability enhancement during ocean transport. The transport vessel does not need to be dry docked to fit or remove stability boxes. See figure 12, which shows how stability boxes improve range of stability only

Figure 12 Stability boxes on the main deck

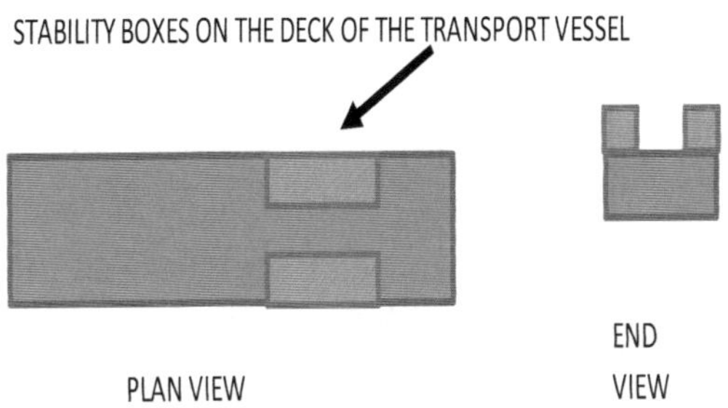

Disadvantages:

- Increased transport vessel weight
- Scrapped after use

4.3 Comparison stability improvers

The sponsons result in a higher initial GM and improved range of intact stability, see figure 13. The transport vessels with Stability boxes have a slightly lower GM than the base vessel, because of the local VCG of the boxes. But the intact range of stability is larger than using sponsons.

Figure 13 Typical GZ curve with stability enhancement

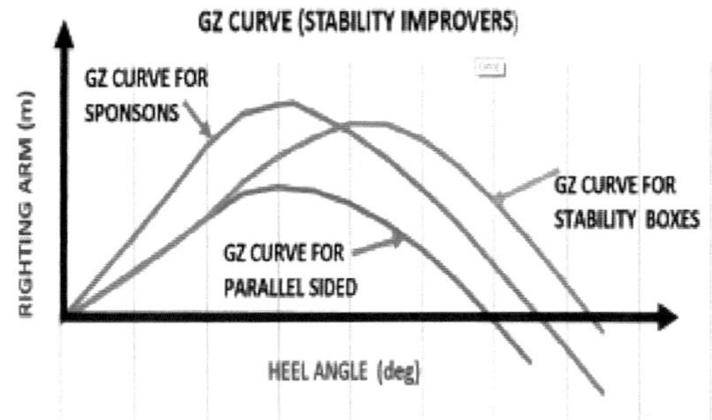

4.4 Stability pontoon

Stability pontoons provide stability enhancement during floating vessel lifting. They are part of the standard equipment for several geared HTVs, see figure 14.
The stability pontoon(s) are fitted and retrieved over the side of the geared HTV vessel using it's own crane.

When a heavy piece of cargo is hanging in the cranes this drastically reduces the stability of the vessel and if it is not sufficiently stable the ship will capsize. A stability pontoon can be connected to the crane vessel during lifting. This makes the vessel able to lift heavier loads or increase the outreach of the cranes.
Stability pontoons on geared HTVs used for inshore sheltered lifting. They would be deployed on one side of the crane vessel and improve initial GM.

Some very large semi submersible crane vessels also use stability pontoons on both sides of the vessel. They form part of the vessel equipment and are deployed using its own cranes, to improve GM

Figure 14 Stability pontoons

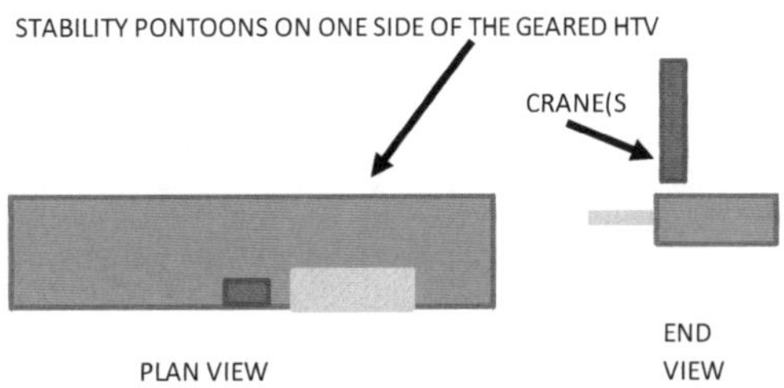

Disadvantages:

- Time taken to deploy
- Time taken to retrieve
- Because the stability pontoon is not the full depth of the vessel then only the initial GM is enhanced
- The stability pontoon can only be deployed in very good weather

Advantages

- The geared HTV with the stability pontoon stored on-board the main deck minimises the transport vessel width during transit.

4.5 Variable shape for different drafts

A few mono hull crane vessels, see figure 15, use dual draft hull design creates an optimal combination between a small waterline breadth during transit and a significantly larger waterline breadth with additional stability during heavy lift operations.

Figure 15 Different drafts-different widths

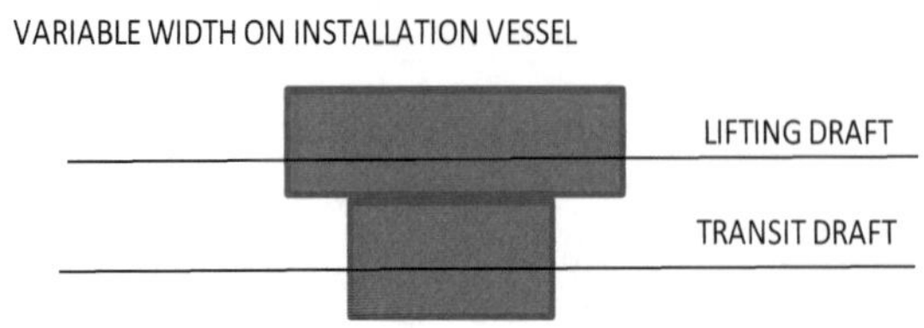

4.6 Review of incidents

Collision of transport vessel during navigation in restricted waters or the open ocean is to be considered. Though for loadout and floatover and offshore lifting this is minimised by having a 500 m safety zone around the operation area.

Grounding of the transport vessel or tug, where a barge is being towed can occur. For example a HTV collided with uncharted submerged rock. Sequential flooding followed by capsize. The Cargo and HTV were scrapped.

Fire or explosion on a towing tug can lead to the towing line being dropped.

Sequential flooding between ballast compartments can occur due to poor maintenance or over loading of a bulkhead due to differential ballasting. For example barge bulkheads corroded at the level of residual ballast. This results in uncontrolled ballasting. Lessons learnt check residual ballast, provide anti-corrosion systems inside ballast tanks.

Wrong ballasting can occur e.g. a HTV sank during offloading of semi submersible, which was undamaged. The HTV was salvaged and repaired. Lesson learnt is to have strict ballasting methods and to only do float-on/float-off operations in shallow water.

Incorrect ballasting can lead to very large heel or trim angles. For example HTV took on a very large heel during loadout. Link beam incorrectly welded, restricting free vertical movement . This resulted in an incorrect ballast sequence, and a large heel angle.

5.0 FUTURE IMPROVEMENTS

Consider developing standard design methods for stability boxes and sponsons. In particular attachment methods for temporary sponsons, which do not require dry docking of the transport vessel need to be developed.

6.0 CONCLUSIONS

The maximum width of an ocean transport vessel is sometimes restricted by the loadout, ocean tow or installation. So temporary stability aids have to be designed in order to meet minimum stability marine warranty requirements.

Damage can occur from collision or stranding but incorrect ballasting or uncontrolled flooding are also possible and hence needs to be considered.

7.0 ACKNOWLEDGEMENTS

The author thanks his colleagues in McDermott for their assistance in preparing this paper. However the paper represents the opinion and views of the author, and does not necessarily represent those of McDermott, or any of its subsidiaries.

8.0 REFERENCES

1. ISO 19901-6:2009 Marine operations
2. DNVGL-ST-N001 Marine operations and marine warranty
3. IMO intact and damage stability requirements Resolution MSC.267(85) (adopted on 4 December 2008)

9.0 AUTHORS BIOGRAPHY

Alan Crowle holds the current position of principal naval architect at McDermott, London. He has a BSc (1974) from University of Newcastle-upon-Tyne and MSc (2020) from the University of Plymouth. He is responsible for naval architecture activities for offshore projects and LNG jetty design. His previous experience includes steel jackets, semisubmersibles, FPSOs, large offshore and onshore modules, and near-shore facilities.

Figure 16 Typical parallel sided cargo barge

Figure 17 Typical Parallel sided launch barge (plus ocean going tug)

Figure 18 Parallel barge with module

Figure 19 T barge used for floatover

DAMAGE PREDICTION ON STIFFENED STRUCTURES BY USING PERIDYNAMICS

C. T. Nguyen and **S. Oterkus,** University of Strathclyde, UK

SUMMARY

Prediction of fracture problems is always challenging work. In order to overcome the difficulties of classical continuum mechanics in terms of using spatial differential equations, the traditional methodologies commonly need additional assumptions for predicting propagation, especially for multiple crack paths. Peridynamics is a new nonlocal theory using integro-differential equations, which are valid in either continuous or discontinuous models. Therefore, peridynamics is suitable for capturing progressive damages. In this study, progressive damages on a plate and a box girder subjected to vertical bending moments are predicted by using peridynamics. Details of the crack locations, crack propagations, as well as the residual strength of the structures, are investigated. A simple method for preventing crack growth is also proposed and investigated.

NOMENCLATURE

$\mathbf{b}$	Vetor of body force (N/m^3)
G_c	Critical energy release rate (J/m^2)
$\mathbf{t}$, $\mathbf{t}'$	Vetor of force densities (N/m^6)
$\mathbf{u}$	Displacement vector (m)
$\ddot{\mathbf{u}}$	Acceleration vector (m/s^2)
V	Volume of a material point (m^3)
$\mathbf{x}$	Coordinate vector (m)
δ	Horizon size (m)
μ	Parameter for representing interaction state
ρ	Mass density (kg/m^3)
φ	Local damage coefficient

1. INTRODUCTION

Structural damage can cause many issues such as losses of human life, environmental pollution, and very expensive maintenance costs. In order to minimize the effect of damages, one of the key tasks is to be able to predict possible progressive damages and understanding the behavior of structures during the damage process.

In the marine industry, structures are commonly made of steel or other ductile metal materials Therefore, ductile fractures are also common damages to these structures. However, in some special conditions, these structures can still experience brittle fractures [1-3].

Conventionally, the finite element method is used for structural analysis. For damage predictions, Peridynamics (PD) is a very suitable method since it is based on integro-differential equations [4-15]. Peridynamics can also be used for analysis beam and shell structures [16-21].

In this study, progressive damages on a plate and a stiffened structure are predicted by using the PD shell model [20]. First, damages on a plate with pre-existing crack subjected to bending are predicted. The plate is also further investigated by assuming some simple methods for preventing crack growth. The residual strength of the plate is also investigated. Next, the strength and damage evolution in a box girder with a cut-out subjected to bending loading conditions is investigated.

2. PERIDYNAMICS

Peridynamics is a new nonlocal theory introduced by Silling [5]. As opposed to classical continuum mechanics, the motion of a material point in PD is influenced by collective deformations of surrounding material points within a domain H_x, which is called horizon. The horizon H_x of a material point can be defined as the circle for a 2D model or a sphere for a 3D model with a radius δ. The radius δ is called horizon size (see Fig. 1). Material points located inside the horizon H_x of a material point are called family members of that material point.

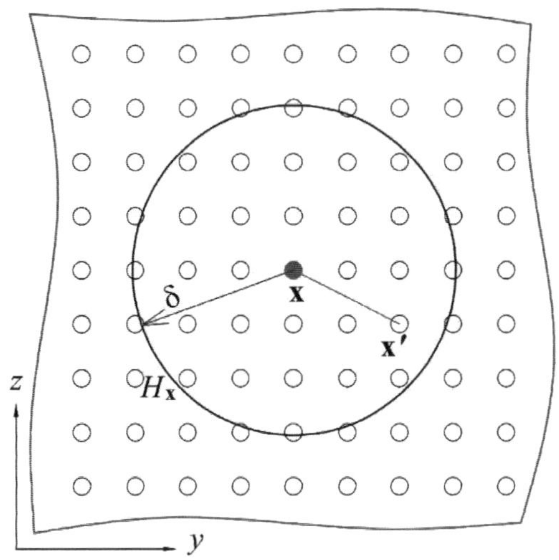

Fig. 1. Peridynamic material points and their family members

In PD, material behaviors are represented by using the integro-differential equation as

$$\rho(\mathbf{x})\ddot{\mathbf{u}}(\mathbf{x},t) = \int_{H_x} \mu(\mathbf{x}'-\mathbf{x},t)\begin{pmatrix}\mathbf{t}(\mathbf{u}'-\mathbf{u},\mathbf{x}'-\mathbf{x},t) \\ -\mathbf{t}'(\mathbf{u}-\mathbf{u}',\mathbf{x}-\mathbf{x}',t)\end{pmatrix} dV' + \mathbf{b}(\mathbf{x},t) \qquad (1)$$

where ρ represents the mass density, $\mathbf{b}$ represents the external body force, V represents the volume of the material point, $\ddot{\mathbf{u}}$ represents the acceleration and $\mathbf{u}$ represents the displacement. The terms $\mathbf{t}$ and $\mathbf{t}'$ represent the force density between the material points, respectively. The term μ represents the failure parameter which can be defined as [4]

$$\mu(\mathbf{x}'-\mathbf{x},t) = \begin{cases} 1 & \text{if interaction exists,} \\ 0 & \text{if no interaction.} \end{cases} \qquad (2)$$

The damage process in PD is simply introduced by eliminating interactions between material points. In order to determine the level of damage on a PD model, the local damage, $\varphi(\mathbf{x}_{(k)},t)$ is defined as [4]

$$\varphi(\mathbf{x},t) = 1 - \frac{\int_H \mu(\mathbf{x}'-\mathbf{x},t)(\mathbf{x}'-\mathbf{x},t)dV'}{\int_H dV'} \qquad (3)$$

which varies between 0 and 1.

In order to determine the value of μ in Eq. (2), a peridynamic damage criterion for each interaction is needed. To date, two commonly used damage criteria are bond stretch criterion [4, 6] and bond energy criterion [9, 10, 19, 20, 22, 23]. In the first criterion, if the bond stretch exceeds the critical value, the bond is considered as irreversibly broken. Similarly, in the second criterion, if the energy release rate of a bond exceeds its critical energy release rate, the bond is also considered as irreversibly broken. The critical stretch and critical energy release rate for a bond are defined from classical material properties.

3. NUMERICAL RESULTS

In this section, damages on shell structures are predicted by using the PD shell model developed by Nguyen and Oterkus [20]. First, progressive damages on a single plate subjected to bending are presented in Section 3.1. Next, progressive damages on a box girder structure subjected to bending are presented in Sec 3.2.

3.1 PLATE SUBJECTED TO BENDING

A plate with dimensions of $L = 0.8$ m, $W = 0.6$ m, and thickness of $h = 0.01$ m is investigated as shown in Fig. 2. The pre-existing crack has a length of $a = W/5$. The material has elastic modulus $E = 2\times10^{11}$ N/m^2 and Poisson's ratio $\nu = 0.27$. The critical energy release rate is $G_c = 2.2714\times10^4$ J/m^2.

The plate is investigated in three cases. Case 1 is the plate with a pre-existing crack a. In case 2, two circular patches are added at two tips of the crack as shown in Fig. 2. The radius of the patch is $r = a/2$. Case 3 is the same as case 2, but with smaller patch radius, $r = a/4$. In both cases, two patches have the same thickness and material properties as the base plate.

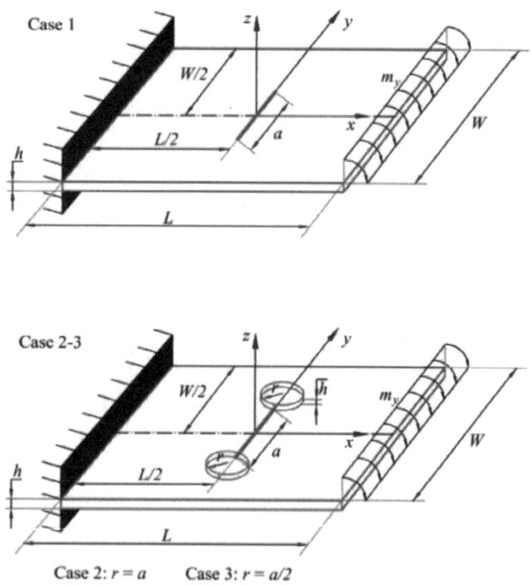

Fig. 2. Plate subjected to bending

The plate is fixed on the left edge and it is subjected bending on the right edge as shown in Fig. 2. In the PD model, the shell is discretized with a mesh size $\Delta x = L/100$. In order to represent the patches, the thickness of the material points located inside the patches is defined as two times thicker than the base plate.

Fig. 3 shows the final damages on the plate in the three cases. The color represents the local damage value φ as given in Eq. (3). As shown in Fig. 3(a) for case 1, the crack path is nearly parallel to the left and right edges of the plate. This observation also agrees very well with the results studied in [20] and [17]. As shown in Fig. 3(b-c) for cases 2 and 3, by putting two patches on the crack tips, the propagation of the initial crack is prevented. However, there are also new damages that appear at the locations which are surrounding the edges of the patches. In the numerical model, these regions are the connection locations between thicker parts and the base plate with normal thickness. It can also be observed that the damages only occur on the connection locations which are near the right edge where the load is applied.

Fig. 4 shows the loading history applied on the plate for the three cases. The moment is calculated as $M = m_y W$. As can be seen from the figure, the maximum bending moment for case 1 is approximately 1440 N.m. After exceeding this peak value, the crack starts growing gradually which leads to the gradually decreased of the required bending moment. At the final damage, the required bending moment reduces to 977 N.m.

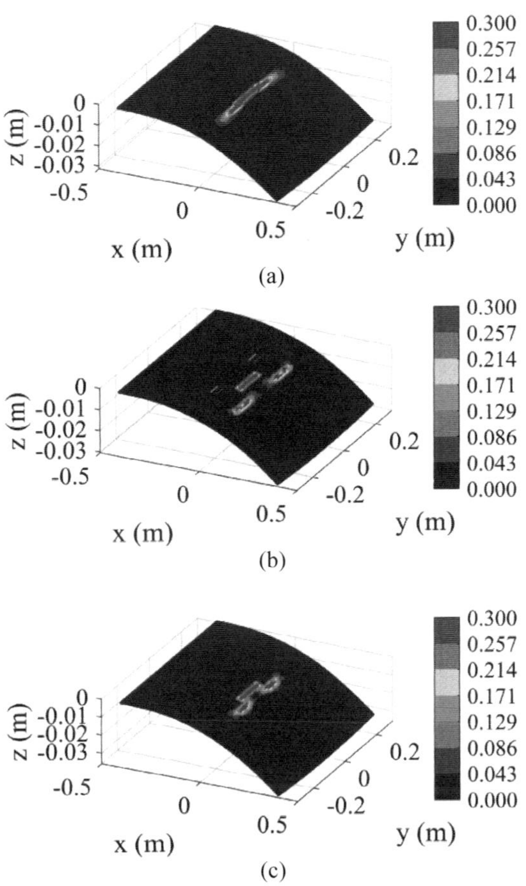

Fig 3. Final damage on the plate in the deformed configurations (a): Case 1, (b): Case 2, (c): Case 3

In cases 2 and 3, because of the contribution of two added patches, the required peak bending moments are much increased compared to case 1. This also means that the strength of the plate is increased. The peak values for bending moments for cases 2 and 3 are 1919 N.m and 2083 N.m, respectively. Therefore, it can also conclude that putting patches on the crack tips helps to reduce the potential of crack growth. For example, if the loading is below 1900 N.m, the crack will not grow in both cases 2 and 3. In addition, it is also interesting to observe that the peak value of applied bending moment in case 3 (with smaller patches) is higher than the value for case 2 (with bigger patches). Therefore, it also can conclude that in order to minimize the potential of crack growth, adding small patches at the crack tips also work well.

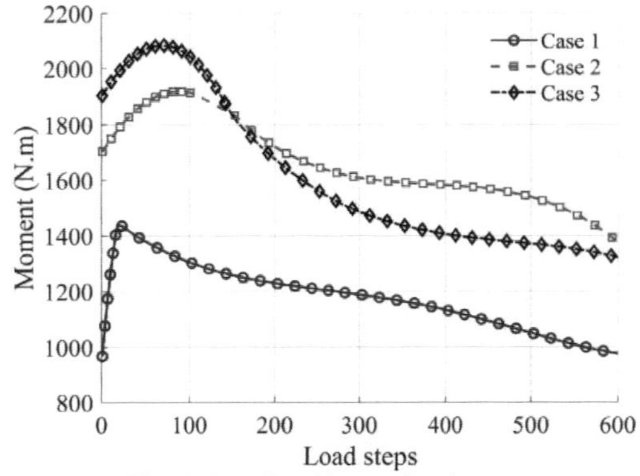

Fig. 4. Loading history for three cases

3.2 A BOX GIRDER SUBJECTED TO BENDING

In this section, damages on a box girder subjected to bending are predicted. The 3D structure and its dimensions are shown in Fig. 5. The dimensions of the structure are $L = 1.44$ m, $B = H = 0.72$ m, $c = 0.05$ m, $d = 0.18$ m. The structure is assumed to have a cut-out at the bottom as shown in Fig. 5. The dimensions of the cut-out are $a = L/3$ and $b = B/3$. The material properties of the structure are the same as the previous section.

In the PD model, the structure is discretized with a uniform mesh size of $\Delta x = 0.02$ m. In order to apply boundary and loading conditions, two ends of the box are attached with two rigid plates. The left rigid plate is fixed, meanwhile, the right rigid plate is subjected to sagging moment, M.

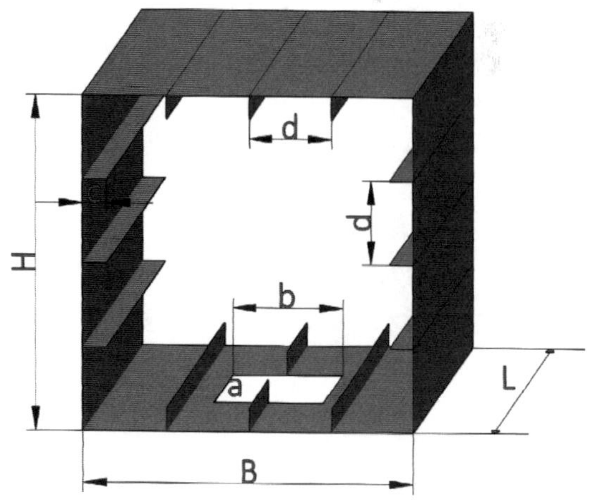

Fig. 5. The geometry of the box girder

Fig. 6-8 show the damage evolution on the structure. As shown in Fig. 6, after 500 load steps, the cracks start from two corners of the cut-out which are near the right end, and propagating transversely and symmetrically to the ship sides. For visualization purposes, the lower parts of the structure with damages are also presented. After 1000 load steps, the cracks propagate vertically beyond the location of the first longitudinal stringers as shown in Fig. 7. After 1340 load steps (final damage), the cracks continue propagating vertically and go beyond the second longitudinal stringers as shown in Fig. 8. Note that, for visualization purposes, the displacements are magnified by 100 times for deformed configurations in Figs. 6-8.

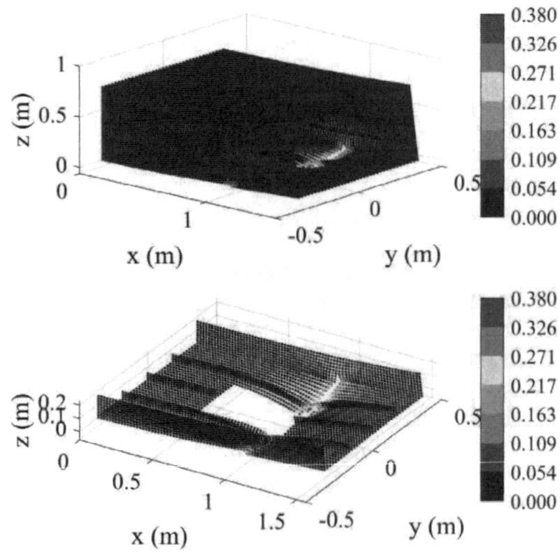

Fig. 6. Damage on the structure after 500 load steps

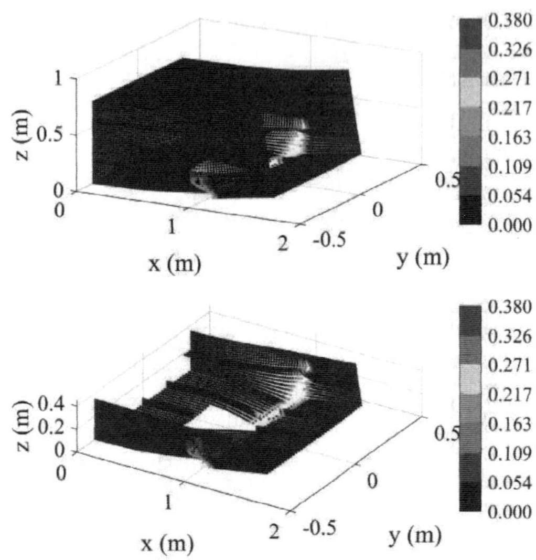

Fig. 7. Damage on the structure after 1000 load steps

Fig. 9 presents the loading history that is applied to the structure. It is obvious that shells on deck and bottom have the main contribution to the bending strength of the structure. Therefore, after 500 load steps, the cracks propagate through the bottom which leads to a significant reduction in the required applied bending moment as shown in Fig. 9. Later, the bending moment reduces gradually to the minimum value of 59083 N.m at the final damage state.

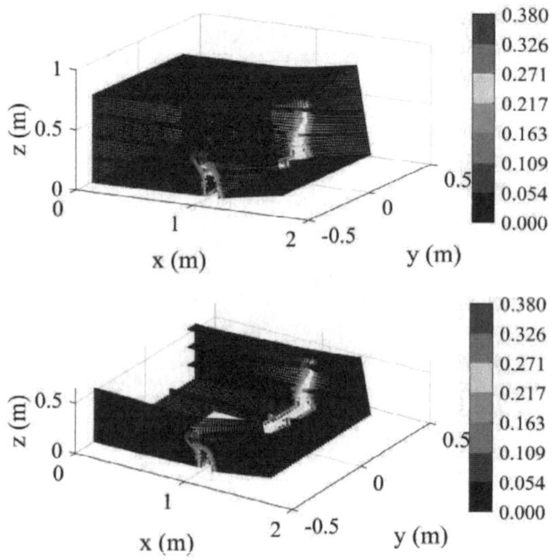

Fig. 8. Final damage on the structure

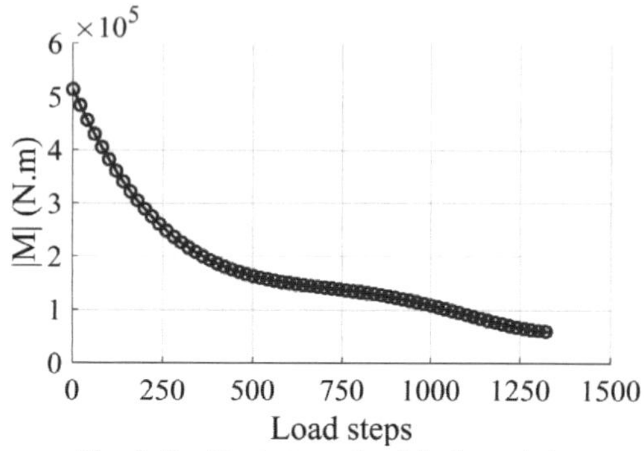

Fig. 9. Residual strength of the box girder

4. CONCLUSION

In this study, progressive damages on stiffened structures are predicted by using the peridynamic shell model. A simple method for preventing the crack growth in a plate with pre-existing crack subjected to bending is proposed. It is found that putting the additional patches at crack tips can help to increase the strength of the structure and it can reduce the possibility of crack growth. However, when the load is critical, new damage locations can still occur at the connection region between thick and thin parts especially for the one near the location of applied loading.

In this study, the effect of putting circular patches is investigated. The other possible methods to prevent crack growth can also be inspected by using the same approach.

The details of damage evolution including locations of crack initiation, crack paths, crack opening, and residual strength of a box girder are investigated by using the peridynamic model. Therefore, further applications of this PD model for damage prediction on ships and offshore structures can be investigated.

5. ACKNOWLEDGEMENTS

The authors gratefully acknowledge the support from the Ministry of Education and Training in Vietnam and the University of Strathclyde. Results were obtained using the ARCHIE-WeSt High Performance Computer (www.archie-west.ac.uk) based at the University of Strathclyde.

6. REFERENCES

1. SUTAR, S.S., G.S. KALE, and S.H. MERAD, *Analysis of ductile-to-brittle transition temperature of mild steel.* Int. J. Innov. Eng. Res. Technol., 2014. **1**: p. 1-10.
2. BENAC, D.J., N. CHEROLIS, and D. WOOD, *Managing cold temperature and brittle fracture hazards in pressure vessels.* Journal of Failure Analysis and Prevention, 2016. **16**(1): p. 55-66.
3. KOBAYASHI, H. and H. ONOUE, *Brittle fracture of liberty ships.* Failure Knowledge Database, 1943. **100**: p. 67.
4. SILLING, S.A. and E. ASKARI, *A meshfree method based on the peridynamic model of solid mechanics.* Computers & structures, 2005. **83**(17-18): p. 1526-1535.
5. SILLING, S.A., *Reformulation of elasticity theory for discontinuities and long-range forces.* Journal of the Mechanics and Physics of Solids, 2000. **48**(1): p. 175-209.
6. MADENCI, E. and E. OTERKUS, *Peridynamic Theory and Its Applications.* Peridynamic Theory and Its Applications. 2014: Springer.
7. FOSTER, J.T., S.A. SILLING, and W.W. CHEN, *Viscoplasticity using peridynamics.* International journal for numerical methods in engineering, 2010. **81**(10): p. 1242-1258.
8. MITCHELL, J.A., *A Nonlocal Ordinary State-Based Plasticity Model for Peridynamics*. 2011, Sandia National Lab.(SNL-NM), Albuquerque, NM (United States).
9. MADENCI, E. and S. OTERKUS, *Ordinary state-based peridynamics for plastic deformation according to von Mises yield criteria with isotropic hardening.* Journal of the Mechanics Physics of Solids, 2016. **86**: p. 192-219.
10. NGUYEN, C.T. and S. OTERKUS, *Ordinary state-based peridynamic model for geometrically nonlinear analysis.* Engineering Fracture Mechanics, 2019.
11. OTERKUS, E., *Peridynamic theory for modeling three-dimensional damage growth in metallic and composite structures*. 2010, The University of Arizona: The University of Arizona.
12. OTERKUS, E. and E. MADENCI. *Peridynamic theory for damage initiation and growth in composite laminate.* in *Key Engineering Materials*. 2012. Trans Tech Publ.
13. ALPAY, S. and E. MADENCI. *Crack growth prediction in fully-coupled thermal and deformation fields using peridynamic theory.* in *54th AIAA/ASME/ASCE/AHS/ASC structures, structural dynamics, and materials conference*. 2013.
14. MADENCI, E. and S. OTERKUS, *Peridynamics for coupled field equations.* Handbook of Peridynamic Modeling, 2016: p. 489-531.
15. OTERKUS, S., *Peridynamics for the solution of multiphysics problems*. 2015, The University of Arizona: The University of Arizona.
16. DIYAROGLU, C., E. OTERKUS, and S. OTERKUS, *An Euler–Bernoulli beam formulation in an ordinary state-based peridynamic framework.* Mathematics Mechanics of Solids, 2019. **24**(2): p. 361-376.
17. DIYAROGLU, C., et al., *Peridynamics for bending of beams and plates with transverse shear deformation.* International Journal of Solids Structures, 2015. **69**: p. 152-168.
18. O'GRADY, J. and J. FOSTER, *Peridynamic beams: a non-ordinary, state-based model.* International Journal of Solids Structures, 2014. **51**(18): p. 3177-3183.
19. NGUYEN, C.T. and S. OTERKUS, *Peridynamics formulation for beam structures to predict damage in offshore structures.* Ocean Engineering, 2019. **173**: p. 244-267.
20. NGUYEN, C.T. and S. OTERKUS, *Peridynamics for the thermomechanical behavior of shell structures.* Engineering Fracture Mechanics, 2019: p. 106623.
21. YANG, Z., et al., *Implementation of peridynamic beam and plate formulations in finite element framework.* Continuum Mechanics Thermodynamics, 2019. **31**(1): p. 301-315.
22. MADENCI, E. and S. OTERKUS, *Ordinary state-based peridynamics for thermoviscoelastic deformation.* Engineering Fracture Mechanics, 2017. **175**: p. 31-45.
23. FOSTER, J.T., S.A. SILLING, and W. CHEN, *An energy based failure criterion for use with peridynamic states.* International Journal for Multiscale Computational Engineering, 2011. **9**(6).

7. AUTHORS BIOGRAPHY

Cong Tien Nguyen is a PhD candidate in Naval Architecture, Ocean and Marine Engineering department at University of Strathclyde. His current research interests include damage prediction of materials and structures by using peridynamics as well as fluid-structure interaction analyses. His PhD research focuses on predicting progressive damages on ships, offshore and marine structures.

Selda Oterkus is an associate professor (senior lecturer) in Naval Architecture, Ocean and Marine Engineering department at University of Strathclyde. Her main research interests are offshore and subsea structures, fluid structure interaction, damage modelling, multiphysics analysis, composite structures, structural health monitoring, corrosion and fatigue. She received her PhD on peridynamics from University of Arizona. She is currently the vice-director of PeriDynamics Research Centre (PDRC) at University of Strathclyde.

ULTIMATE STRENGTH PERFORMANCE OF A DAMAGED CONTAINER SHIP

S. Li, Z.Q. Hu & S.D. Benson, Newcastle University, UK

SUMMARY

The ultimate strength performance is assessed of a container ship with side-shell damage. Simplified progressive collapse method is employed for predicting the ultimate ship hull strength under vertical bending. A parametric study is conducted to identify the most critical damaged scenario. Equivalent nonlinear finite element analysis is performed for validation. The computed strength is compared with the extreme design bending moment for a safety margin assessment. Residual strength index versus damage extent diagrams are proposed, which may be useful for a rapid prediction of the residual ultimate strength for container ships of similar class under a given damaged scenario.

NOMENCLATURE

M_u	Ultimate bending strength
M_p	Plastic collapse strength
Φ	Critical stiffened panel strength
ΔM_H	Horizontal bending moment increment
ΔM_V	Vertical bending moment increment
$\Delta \chi_H$	Horizontal curvature increment
$\Delta \chi_V$	Vertical curvature increment
z_G	Vertical centroid of neutral axis
y_G	Horizontal centroid of neutral axis
w_p	Local plate deflection
v_s	Stiffener sideway deflection
w_c	Column-type deflection
a	Plate length
b	Plate width
t	Plate thickness
w_c	Web height
β	Plate slenderness ratio
λ	Column slenderness ratio

1. INTRODUCTION

The continuing development of container shipping industry in recent years gives rise to the concern on container ship structural safety. During the service period of container ships, ship-ship collision may occur in some of the busiest sea routes and canals due to various unexpected operational errors. For instance, two container ships collided in Suez Canal in 2014. No causality was reported as a result of this accident, but a 65 feet long dent was left on one of the container ships. More recently, a collision took place between a ro-ro ferry and a container ship in the Mediterranean Sea, causing breach of side shell of the container ship. Generally, these accidents will lead to significant structural damages on the side shell of the vessels, which can reduce their safety levels in the remaining journey. A rapid evaluation on whether the damaged ship should be immediately tugged to the nearest port for repair or it can continue to complete the scheduled voyage is therefore highly important.

As a critical indicator of the structural performance, the residual ultimate ship hull strength is highly important to assess the consequence of a damaged scenario. Dow [1] investigated the structural redundancy and damage tolerance of ship hull structures in relation to the ultimate strength performance using a destroyer model. Various design features that may result in an improve of the ultimate ship hull girder were discussed. It was indicated that the column slenderness ratio of stiffened panel should never exceed 0.55 to achieve an imperfection–insensitive cross section design. The influence of structural damage to the ultimate ship hull strength was analysed on a bulk carrier and tanker by Fujikubo et al [2]. In terms of the residual ultimate ship hull strength, it was found that the effect of neutral axis rotation may not be important when only the outer side shell plating was damaged. However, the difference occurred in the post-collapse response where a drastic post-collapse drop in strength tool place if the neutral axis rotation was accounted for. A comparison of numerical methods to predict the ultimate strength of a damaged box girder was completed by Benson et al [3]. The use of static finite element method and dynamic finite element method were compared. It was concluded that either approach was suitable for analysing the progressive collapse behaviour of box girder type structures. The neglection of mass and damping effects are acceptable even though buckling is in fact a dynamic phenomenon. Meanwhile, the comparative study also demonstrated that the simplified progressive collapse method can provide closely correlated results in comparison to a more computationally expensive finite element method.

Regarding to damaged container ships, the residual ship hull strength was assessed by Kim et al [4], in which different grounding damaged scenarios were considered and a series of grounding damage index diagrams were proposed. Tekgoz et al [5] conducted a strength assessment for a container ship with side shell damage subjected to asymmetric bending. The study suggested that the finite element model extent had a negligible effect on the ultimate ship hull strength, but the post-collapse range was considerably affected.

Comparing with other ship types, the investigation is generally scarce on the ultimate strength performance of damaged container ship. In this regard, this paper presents an evaluation of the residual ultimate strength of a container ship with side shell damage. A variety of damage scenarios with different damage extents and locations are analysed, such that the most critical damaged case can be identified. The ultimate ship hull girder is calculated using a simplified progressive collapse method for vertical bending where the effect of neutral axis rotation is taken into account. The computed strength is compared with the extreme design bending moment for a safety margin assessment. A series of residual strength index (RSI) versus damage extent diagrams are established, which might be useful for a rapid estimation of the residual ship hull strength for similar class container ships for a given side shell damaged scenario.

2. BACKGROUND

As the most fundamental safety assessment for ship structures, a large body of methodology has been developed to predict the ultimate hull girder strength. Generally, the ultimate hull girder strength can be evaluated using one of the following methods:

- Empirical formulae
- Presumed stress distribution-based method
- Simplified progressive collapse method
- Idealised structural unit method
- Nonlinear finite element method

2.1 EMPIRICAL FORMULAE

An empirical formula was proposed by Frieze and Lin [6] based on the numerical results on a series of hull girder. It was indicated by Frieze and Lin that the ultimate moment capacity of longitudinally framed hulls under vertical bending closely correlated with the ultimate strength of the critical compression panel. Therefore, an empirical relationship between the ultimate hull girder strength and critical stiffened panel strength was proposed taking form defined by Equation (1). The coefficients were determined from regression analysis. Different sets of coefficients were derived for sagging and hogging (Table 1), as there are many hard corners provided by the keel plates, deep girders and bilge keels which exist in the compressed part of a vessel in hogging.

$$M_u/M_p = d_1 + d_2\Phi + d_3\Phi^2 \tag{1}$$

Table 1. Coefficients of Frieze and Lin's formulae

	d_1	d_2	d_3
Sagging	-0.172	1.548	-0.368
Hogging	0.003	1.459	-0.461

2.2 PRESUMED STRESS DISTRIBUTION-BASED METHOD

The presumed stress distribution-based method was originally introduced by Caldwell [7]. Fundamentally, a stress distribution was assumed at the collapse state of the cross section and the ultimate bending moment was calculated by taking the second moment of the stress distribution. The core of this methodology is the assumed stress distribution. In the original Caldwell's formulation, a uniform stress distribution was employed (Figure 1a). However, at the state of collapse, the side shell of the cross section usually remains elastic, especially for those near the neutral axis. Therefore Caldwell's formulation may lead to an overestimation of the ultimate hull girder strength. In this regard, Paik and Mansour [8] proposed a different stress distribution for the cross section at collapse state (Figure 1b). The validation presented by Paik and Mansour [8] has shown the improved accuracy of their method compared with Caldwell's original formulation.

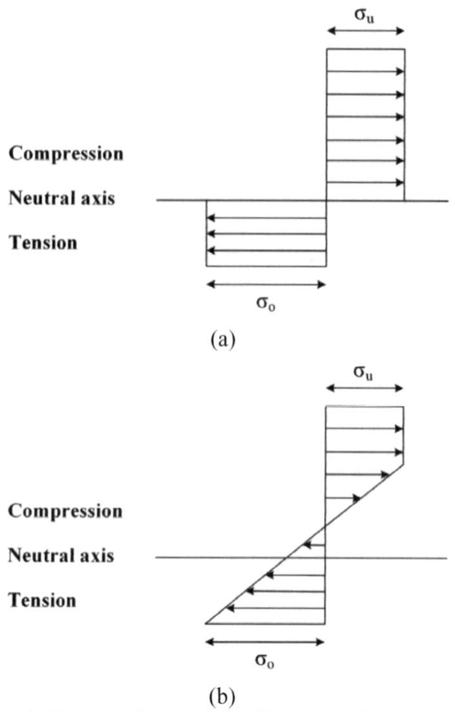

Figure 1. Presumed stress distribution of hull girder at sagging collapse; (a) Caldwell's original stress distribution (1965); (b) Paik and Mansour's modified stress distribution.

2.3 SIMPLIFIED PROGRESSIVE COLLAPSE METHOD

Although the ultimate hull girder bending moment can be predicted, the empirical formulae and presumed stress distribution-based method are not able to predict the overall bending moment/curvature relation of the hull girder from pre-collapse range to post-collapse range. This issue was resolved by Smith [9] with a simplified progressive collapse method. A simple analysis procedure was proposed by Smith as follows:

(i) The hull cross section is divided into small elements as shown in Figure 2;
(ii) Vertical curvature of the hull is assumed to occur incrementally; the corresponding incremental element strains are calculated on the assumption that plane sections remain plane and that bending occurs about the instantaneous neutral axis of the cross section;

(iii) Element incremental stress are derived from incremental strains using the slopes of stress-strain curves;

(iv) Element stresses are integrated over the cross section to obtain bending moment increments; incremental curvatures and bending moments are summed to provide the cumulative values.

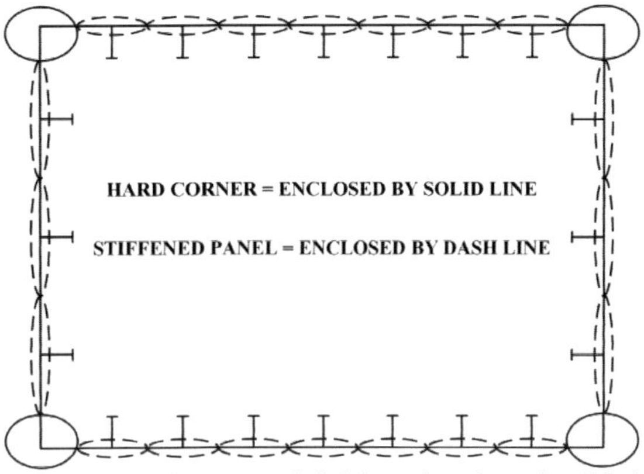

Figure 2. Element subdivision in the simplified progressive collapse method

This simple analysis procedure has been refined in an incremental form (Equation 2) and extended to deal with bi-axial bending problem [10].

$$\begin{Bmatrix} \Delta M_H \\ \Delta M_V \end{Bmatrix} = \begin{bmatrix} D_{HH} & D_{HV} \\ D_{VH} & D_{VV} \end{bmatrix} \begin{Bmatrix} \Delta \chi_H \\ \Delta \chi_V \end{Bmatrix} \tag{2}$$

where ΔM_H, ΔM_V, $\Delta \chi_H$ and $\Delta \chi_V$ are horizontal bending increment, vertical bending increment, horizontal curvature increment and vertical curvature increment respectively. D_{HH}, D_{VV}, D_{HV}, and D_{VH} are tangent rigidities of cross section, defined by Equation (3) to (5).

$$D_{HH} = \sum_{i=1}^{n} k_i A_i (y_i - y_G)^2 \tag{3}$$

$$D_{VV} = \sum_{i=1}^{n} k_i A_i (z_i - z_G)^2 \tag{4}$$

$$D_{HV} = D_{VH} = \sum_{i=1}^{n} k_i A_i (y_i - y_G)(z_i - z_G) \tag{5}$$

where k_i is the effective tangent modulus of ith element, obtained from the slope of a stress-strain curve. A_i is the element cross sectional area assuming that it is small enough for own inertia to be negligible. y_i and z_i are horizontal and vertical ordinates of the element. The location of centroidal axes (y_G, z_G) changes progressively as a result of elastoplastic buckling in parts of the cross section.

2.4 IDEALISED STRUCTURAL UNIT METHOD

In parallel to the development of Smith-type simplified progressive collapse method, an idealised structural unit method (ISUM) was proposed by Ueda and Rashed [11]. ISUM follows the general framework of the nonlinear finite element method, but allows for the use of larger element size, which is attributed to the simplification of the evaluation of geometric and material nonlinearity. With ISUM element, the number of degree of freedom can be considerably reduced compared with the conventional finite element modelling.

2.5 NONLINEAR FINITE ELEMENT METHOD

The application of nonlinear finite element method (NLFEM) to perform collapse analysis on ship structures was initiated by Chen [12]. With an increasingly improved computer power, the nonlinear finite element has become a commonly used approach in this field. For example, the NLFEM was applied to analyse the progressive collapse of bulk carrier in alternate hold loading condition in [13] The investigation on the MOL Comfort accident was conducted by applying NLFEM to predict the ultimate collapse strength [14]. A concise introduction to the NLFEM for ultimate strength assessment is given in [15]. Practical finite element modelling techniques for ship structures can be found in [16].

3. METHODOLOGY

The simplified progressive collapse method is adopted to predict the ultimate ship hull strength. For a symmetric intact cross section, the application of vertical curvature will induce pure vertical bending moment and the neutral axis only progressively translates in the vertical direction. However, an asymmetrically damaged cross section will lead to a rotation of the neutral axis (Figure 3). Hence the analysis should be taken as a bi-axial bending problem. Fujikubo et al. [2] proposed an adapted bending moment/curvature equation (Equation 6 and 7). Two different cases were considered, namely pure bending and constraint bending. The former case considers the effect of neutral axis rotation while it is omitted in the latter one.

Pure vertical bending:

$$\begin{Bmatrix} 0 \\ \Delta M_V \end{Bmatrix} = \begin{bmatrix} D_{HH} & D_{HV} \\ D_{VH} & D_{VV} \end{bmatrix} \begin{Bmatrix} \Delta \chi_H \\ \Delta \chi_V^o \end{Bmatrix} \tag{6}$$

$$\Delta \chi_H = -\frac{D_{HV}}{D_{HH}} \Delta \chi_V^o$$

$$\Delta M_V = \left(D_{VV} - \frac{D_{VH} D_{HV}}{D_{HH}} \right) \Delta \chi_V^o$$

Constrained vertical bending:

$$\begin{Bmatrix} \Delta M_H \\ \Delta M_V \end{Bmatrix} = \begin{bmatrix} D_{HH} & D_{HV} \\ D_{VH} & D_{VV} \end{bmatrix} \begin{Bmatrix} 0 \\ \Delta \chi_V^o \end{Bmatrix} \tag{7}$$

$$\Delta M_H = -D_{HV} \Delta \chi_V^o$$

$$\Delta M_V = D_{VV} \Delta \chi_V^o$$

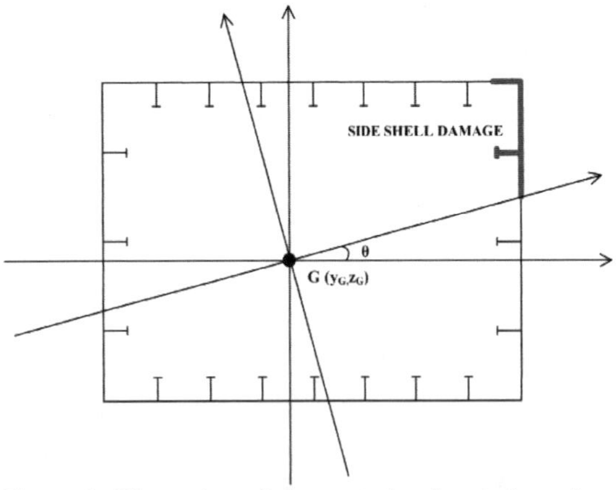

Figure 3. Illustration of the neutral axis rotation of an asymmetrically damaged cross section

The structural components load-shortening curves is predicted by two different formulations, namely an in-house method and the Common Structural Rule (CSR) method. The in-house method is based on elastic large deflection analysis (ELDA) and rigid plastic mechanism analysis (RPMA). ELDA and RPMA are performed independently for the local plate, stiffener web and flange of a stiffened panel. The ultimate strength is determined as the intersection of the solutions of ELDA and RPMA. The pre-ultimate load-shortening response is described by the ELDA solution while the post-ultimate response is described by the RPMA prediction.

Nonlinear finite element method is utilised for a validation of the simplified method calculation. The finite element modelling for elastoplastic buckling collapse analysis requires the consideration of initial imperfection of the structures. The initial imperfection of ship structures are normally developed in the form of initial deflection and residual stress. The modelling of initial deflection can be achieved by a direct node translation technique where the coordinate of each node is offset with respect to the perfect structure in accordance with the prescribed deflection shape. Usually, there are three types of initial deflection being consideration for stiffened panels, namely local plate deflection w_p, beam-column deflection and stiffener sideway deflection (Figure 4). Equation (8) to (10) give the expression of commonly used initial deflection shape. The effect of fabrication-induced residual stress is not considered in the present study.

$$\frac{w_p}{w_{op}} = \left\{0.8sin\left(\frac{\pi x}{a}\right) + 0.2sin\left(\frac{m\pi x}{a}\right)\right.$$

$$\left. +0.01sin\left[\frac{(m+1)\pi x}{a}\right]\right\} sin\left(\frac{\pi y}{b}\right) \quad (8)$$

$$m = int\left(\frac{a}{b}\right) + 1$$

$$w_{op} = \min\left(6, 0.1\beta^2 t\right)$$

$$\frac{v_s}{v_{os}} = \frac{z}{h_w}\left(0.8sin\left(\frac{\pi x}{a}\right) + 0.2sin\left(\frac{i\pi x}{a}\right)\right) \quad (9)$$

$$i = int\left(\frac{a}{h_w}\right) + 1$$

$$\frac{w_c}{w_{oc}} = sin\left(\frac{\pi x}{a}\right) sin\left(\frac{\pi y}{B}\right) \quad (10)$$

$$w_{oc} = \begin{cases} 0.0008a, & \lambda < 0.2 \\ 0.0012a, & 0.2 < \lambda < 0.6 \\ 0.0015a, & \lambda \geq 0.6 \end{cases}$$

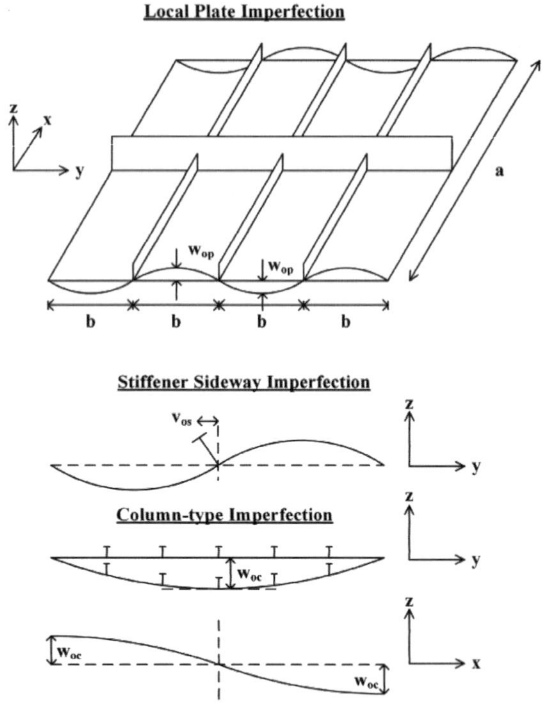

Figure 4. Illustration of the initial distortion of stiffened panels

4. CASE STUDY

4.1 MODEL CHARACTERISTICS

The residual ultimate ship hull strength and safety margin of a 10000TEU container ship is examined in the present study. The principal dimensions of the vessel is summarised in Table 2. The general structural configuration of the case study cross section is shown in Figure 5 and the detailed scantlings are given in the Appendix.

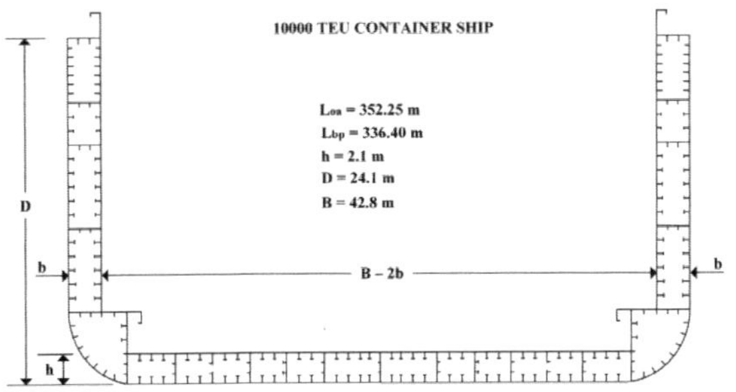

Figure 5. Midship cross section of the case study 10000TEU container ship

As reported by Mohammed et al. [17], the distribution of still water bending moment is illustrated in Figure 6 with the maximum still water load and long-term prediction of extreme wave load summarised in Table 3.

Table 2. Principal dimension of the case study 10000TEU container ship

Principal particular	Dimension
L_{oa}	352.25 m
L_{bp}	336.40 m
h	2.1 m
D	24.1 m
B	42.8 m
Frame spacing	791 mm

Table 3. Maximum still water vertical bending moment and wave-induced vertical bending moment [17]

Item	Vertical BM (GNm)
Max. still water	5.69
Midship still water	4.46
Wave-induced	13.66

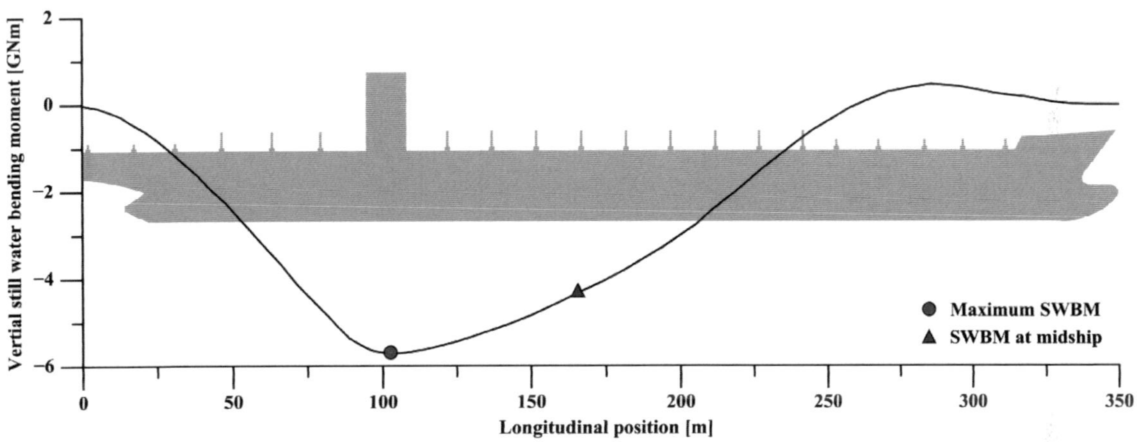

Figure 6. Vertical still water bending moment distribution of the case study 10000TEU container ship

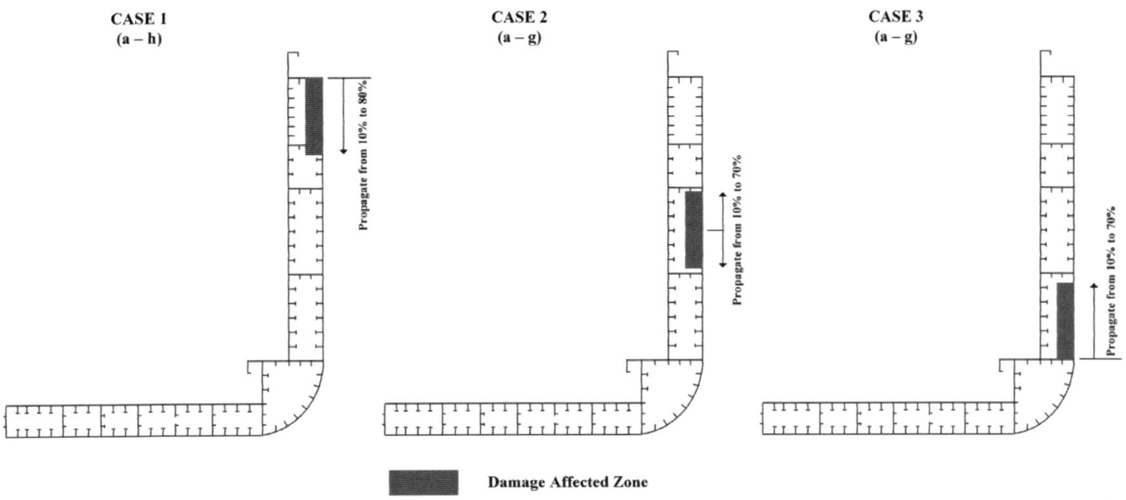

Figure 7. Damage scenarios with different extents and centroid locations

4.2 SCOPE OF ANALYSIS

The present study confines to the assessment of vertical bending. As illustrated in Figure 7 and summarised in Table 4, a range of damage scenarios with different damage extents and centroid locations are analyses. In general, three major groups are analysed, in which the vertical extent of the damaged zone is extended from the deck (Case 1), the centre of the side shell (Case 2) and the inner bottom (Case 3) respectively. For each case, the vertical extent is varied from 0.1z/D to 0.8z/D and the transverse extent of each damaged scenario is 0.5y/b where z represents the vertical damage extent and y corresponds to the transverse damage extent. As the damaged extent is the same in the case of 0.8z/D for each group, the number of damaged scenarios is 22 in total.

Table 4. Damaged scenarios

ID	z/D	y/b
1a~h	10% ~ 80%	0.5
2a~g	10% ~ 70%	0.5
3a~g	10% ~ 70%	0.5

Nonlinear finite element analysis is performed for the Case 1(a) to Case 1(d). A nine-bay FE model is adopted where the damage affected zone is represented by removing the corresponding element at the middle three bays as illustrated in Figure 8. As suggested by Tekgoz et al (2018), the overall finite element model extent had a negligible effect on the ultimate ship hull strength prediction. Regarding to the longitudinal extent of damaged affected zone, its influence might also be insignificant for the present structural configuration and the vertical bending loading case.

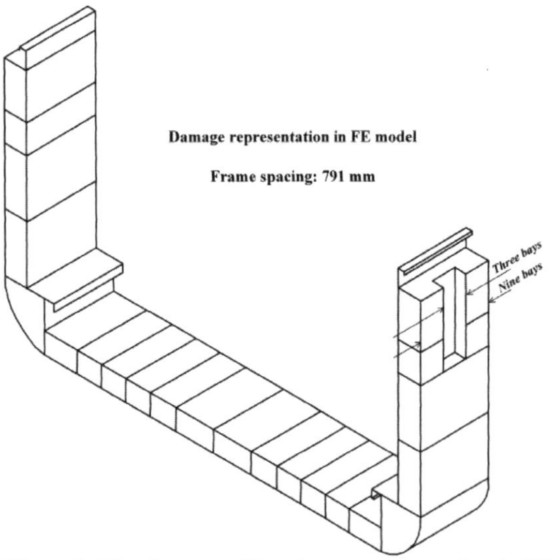

Figure 8. The damage affected zone representaion in FE model

4.3 RESULTS AND DISCUSSIONS

4.3 (a) Progressive Collapse Behaviour

A comparison of the predicted vertical bending moment/curvature relationships by different methods are shown in Figure 9 to 12, in which NLFEM representing the nonlinear finite element method, ProColl indicating the in-house formulation and CSR corresponding the Common Structural Rule approach. In each graph, the results for intact case are included for comparison.

A higher prediction of ultimate ship hull strength and the elastic bending stiffness by NLFEM is shown. This may be attributed to the difference in the assumed imperfection characteristics. In an intact status, a relatively considerable post-collapse drop in strength can be found when the ship is subjected to hogging load, whereas the cross section can still resist a large bending moment after the collapse in sagging. This feature of sagging collapse behaviour is observed in the damaged cross section. However, being similar to the typical response in sagging , a different collapse behaviour in hogging appears for the hull girder with damage,.

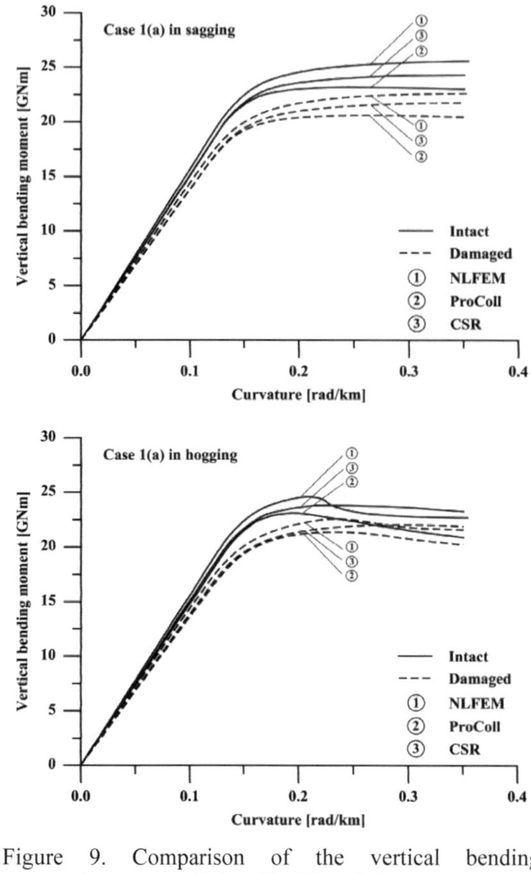

Figure 9. Comparison of the vertical bending moment/curvature relationship (Case 1a)

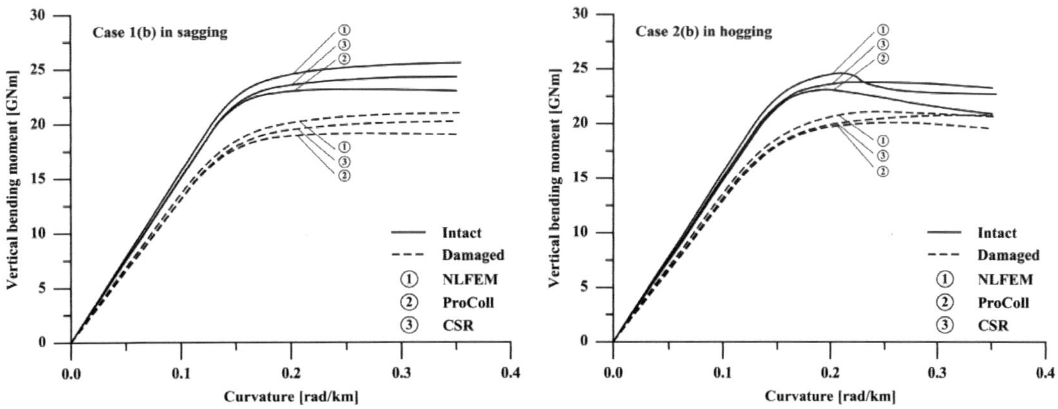

Figure 10. Comparison of the vertical bending moment/curvature relationship (Case 1b)

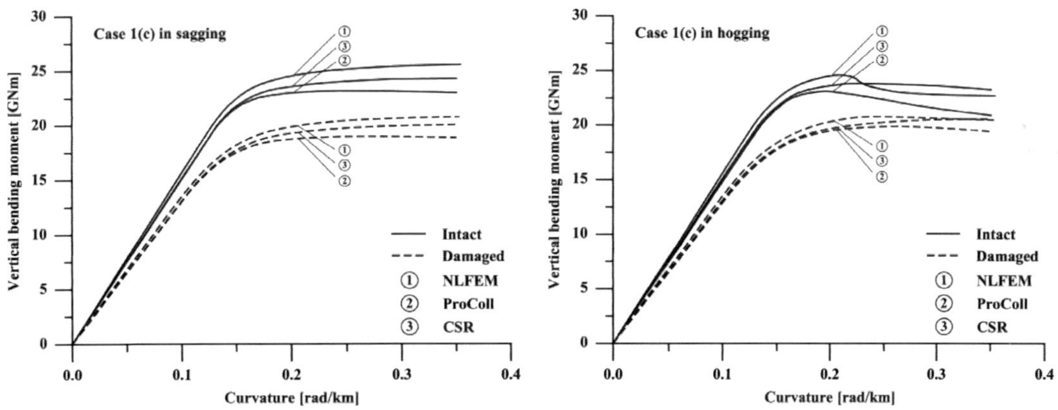

Figure 11. Comparison of the vertical bending moment/curvature relationship (Case 1c)

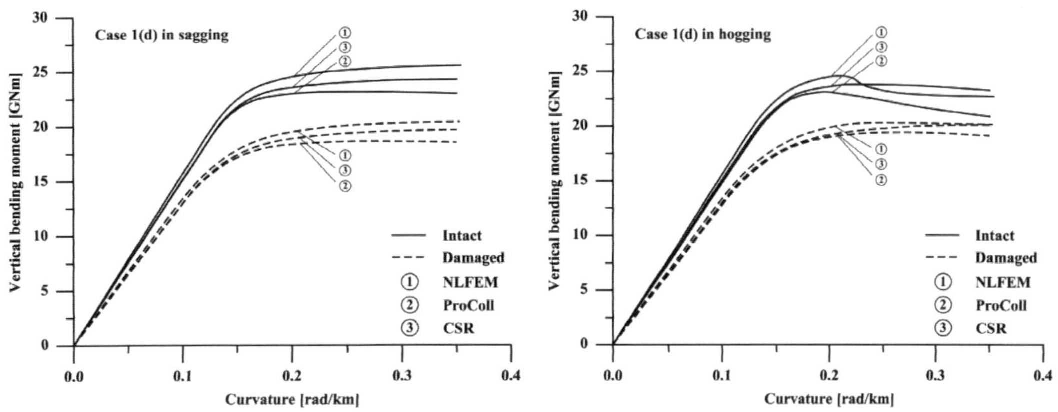

Figure 12. Comparison of the vertical bending moment/curvature relationship (Case 1d)

The collapse deformation and stress distribution predicted by the NLFEM is shown in Figure 13 and 14. At sagging collapse state, the torsion box experiences a gross yielding failure. Buckling nucleation is observed in the inner and outter side shell panels. For Case 1(a), the deformation nucleates at the middle bay, whereas nucleation occurs at the adjacent bay for Case 1(b) to 1(d). Meanwhile, due to the rotation of neutral axis, yielding occurs at the undamaged side of the bottom panels, whereas the bottom panels at the damaged side remains elastic. At hogging collapse, a gross yiedling failure is obserbed at the side shell panels. The buckling deformation occurs at the undamaged side of the bottom panel and nucleates to the transverse bay near the model boundary.

Damaged Scenarios Case 1a to Case 1d in Sagging

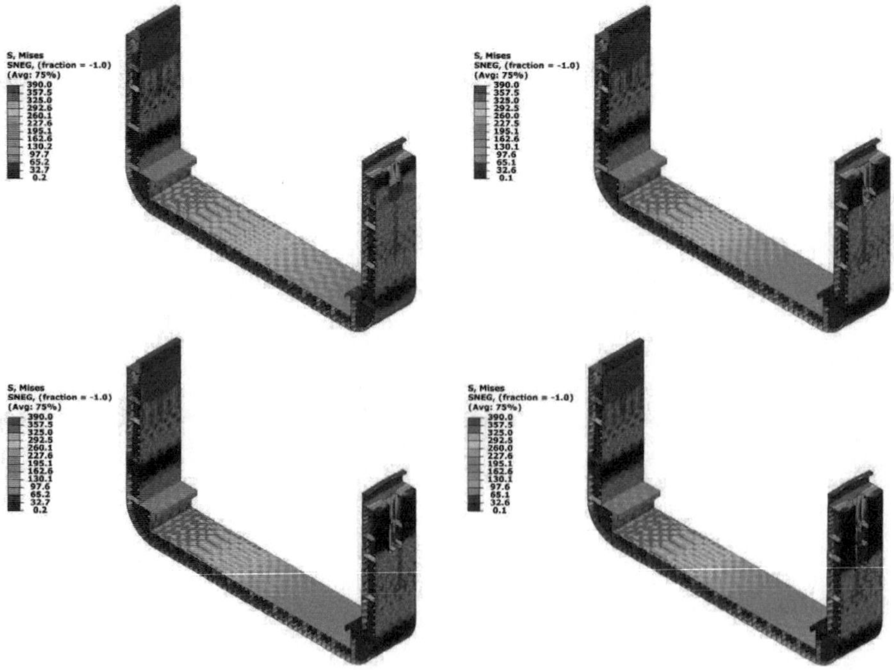

Figure 13. Collapse deformation and stress distribution at sagging predicted by NLFEM

Damaged Scenarios Case 1a to Case 1d in Hogging

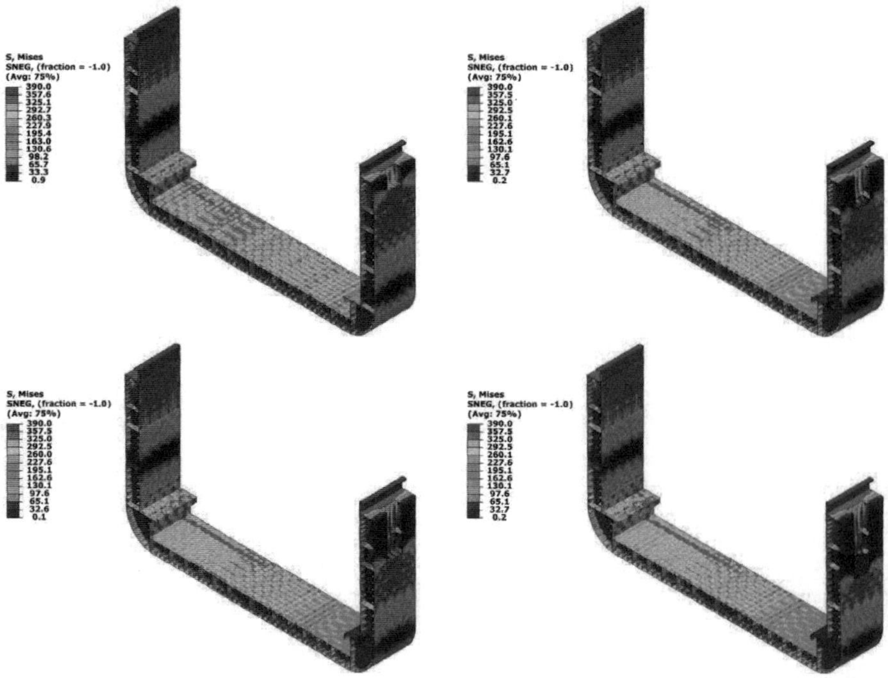

Figure 14. Collapse deformation and stress distribution at hogging predicted by NLFEM

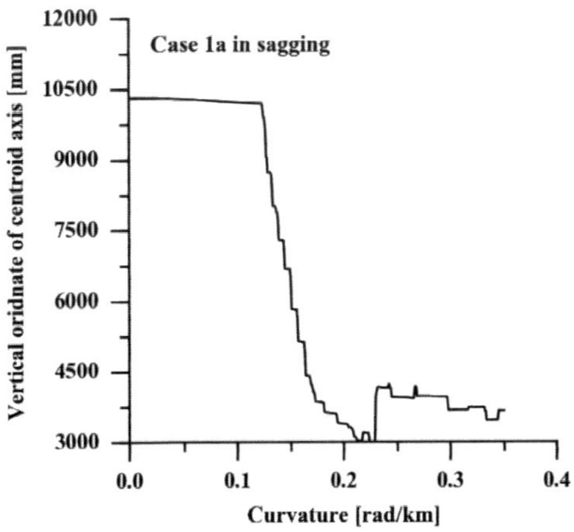

(a)

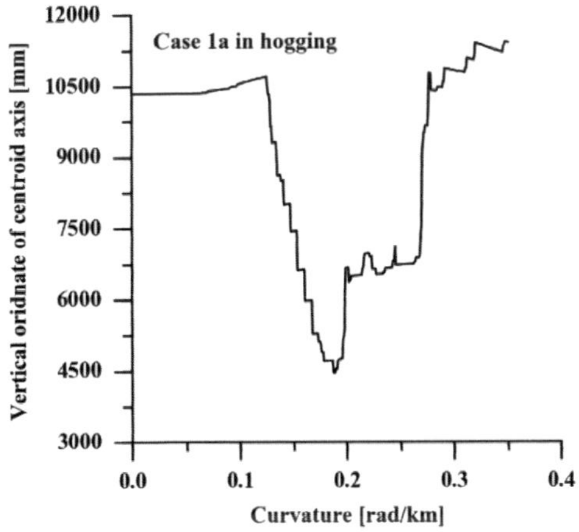

(b)

Figure 15. Translation of the instantaneous neutral axis during progressive collapse

The typical progressive collapse of container ship may be further elucidated with reference to the translation of neutral axis position shown in Figure 15. In sagging, the neutral axis slightly translate toward the bottom until collapse occurs at the upper of the cross section which results in a rapid drop of the neutral axis. With further loading application, the neutral axis experiences an upward translation due the yielding at the outer bottom. In hogging, the neutral axis starts with a slow translation toward to the upper part of the cross section. Unlike bulk carriers and oil tankers, the initial failure takes place at the upper part of container ships under hogging, which therefore leads to the rapid downward translation of the neutral axis. Upon the collapse takes place at the bottom, the neutral axis recovers to be closed to the initial position. This collapse behaviour is attributed to the fact that stocky panels with a small column slenderness ratio are employed in the container ship cross section, which leads to a high compressive ultimate strength close to material yield stress and an insignificant drop of post-collapse strength. Because of the large opening at deck, the initial neutral axis position is relatively closer to the bottom and therefore the resultant stress at a given load is higher at the deck. Thus, failure would first take place at the deck even in the case of hogging.

4.3 (b) Safety Margin Evaluation

The reserve safety margin as defined by Equation (11) is shown in Figure 16 as a function of the vertical damage extent. A significantly lower safety margin is shown for the damaged scenario Case 1 even when the damaged extent is 10%. On the other hand, it is clear that except for a fully damage of the side shell panels, the residual safety margin of Case 2 and 3 is nearly unaffected, in which case the damage affected zone has not propagated to the torsion box. Conversely, the safety margin may be inadequate for Case 1 with a vertical damage extent large than 50%, as indicated by the strength calculation using the in-house formulation (ProColl). Based on this safety margin evaluation, it might be concluded that the damage of torsion box has a dominating detrimental impact on vertical bending strength of a container ship and an adequate safety margin can generally retain if it is still intact.

$$\text{Safety margin} = \left(\frac{M_u - M_d}{M_d}\right) \times 100\% \qquad (11)$$

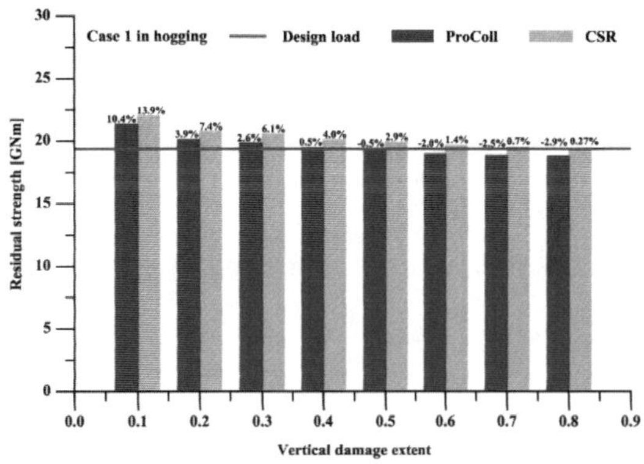

(a)

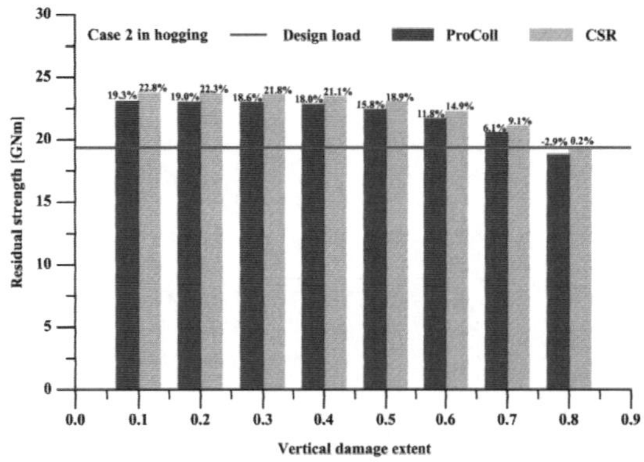

(b)

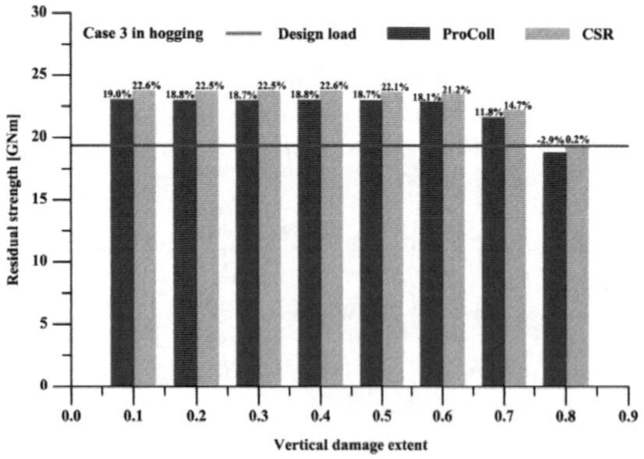

(c)

Figure 16. Residual safety margin of the container ship

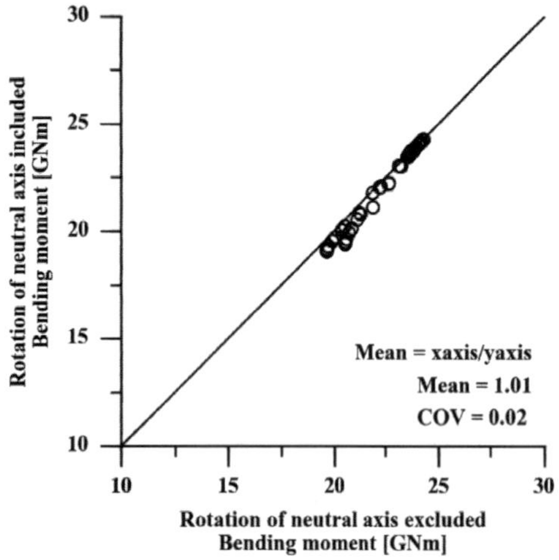

Figure 17. Comparison of the ultimate ship hull strength with and without consideration of the neutral axis rotation

4.3 (c) Effect of Neutral Axis Rotation

A comparison of the calculated ultimate ship hull strength with and without considering the influence of neutral axis rotation, i.e. pure vertical bending and constrained vertical bending, is shown in Figure 17. As indicated by the statistical analysis, the results are in close correlation with a minor reduction in strength when the neutral axis rotation is taken into account in the simplified progressive collapse method calculation. This comparison is in agreement with the observation from the collapse analysis of bulk carrier and oil tankers conducted by Fujikubo et al. Thus, the conclusion may also be valid on the container ships that the neutral axis rotation has an insignificant influence on the prediction of residual ship hull strength if only the outer side shell is damaged.

4.3 (d) Residual Strength Index (RSI)

The calculation results using simplified progressive collapse method are expressed in the graphical form where the residual strength index (RSI) is given in terms of the vertical damaged extent (Figure 18). The RSI is defined by Equation (7) where $M_u^{damaged}$ indicates the residual ultimate strength of a damaged structure and M_u^{intact} represents the ultimate strength of an intact structure. These diagrams may provide a rapid evaluation of the residual ship hull strength of container ships of similar class. It can also be observed from these diagrams that the reduction in ultimate strength predicted by the in-house formulation and CSR method is highly correlated.

$$\text{Residual strength index} = \frac{M_u^{damaged}}{M_u^{intact}} \qquad (12)$$

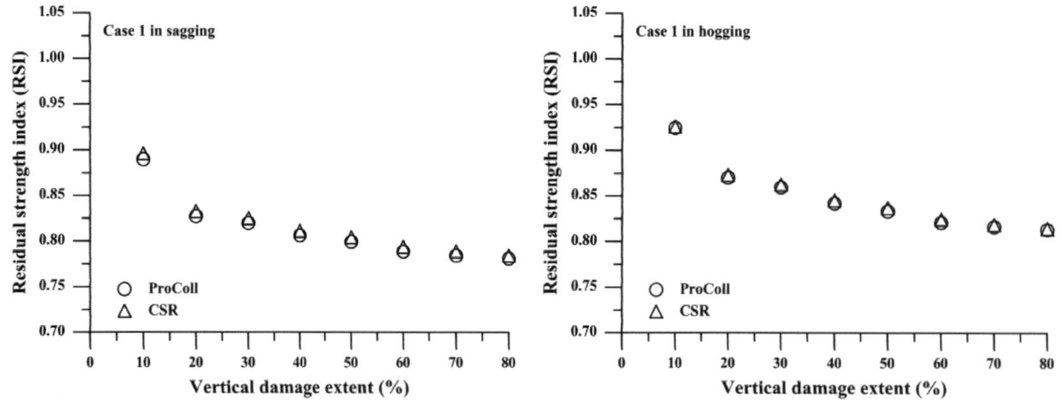

(a) (b)

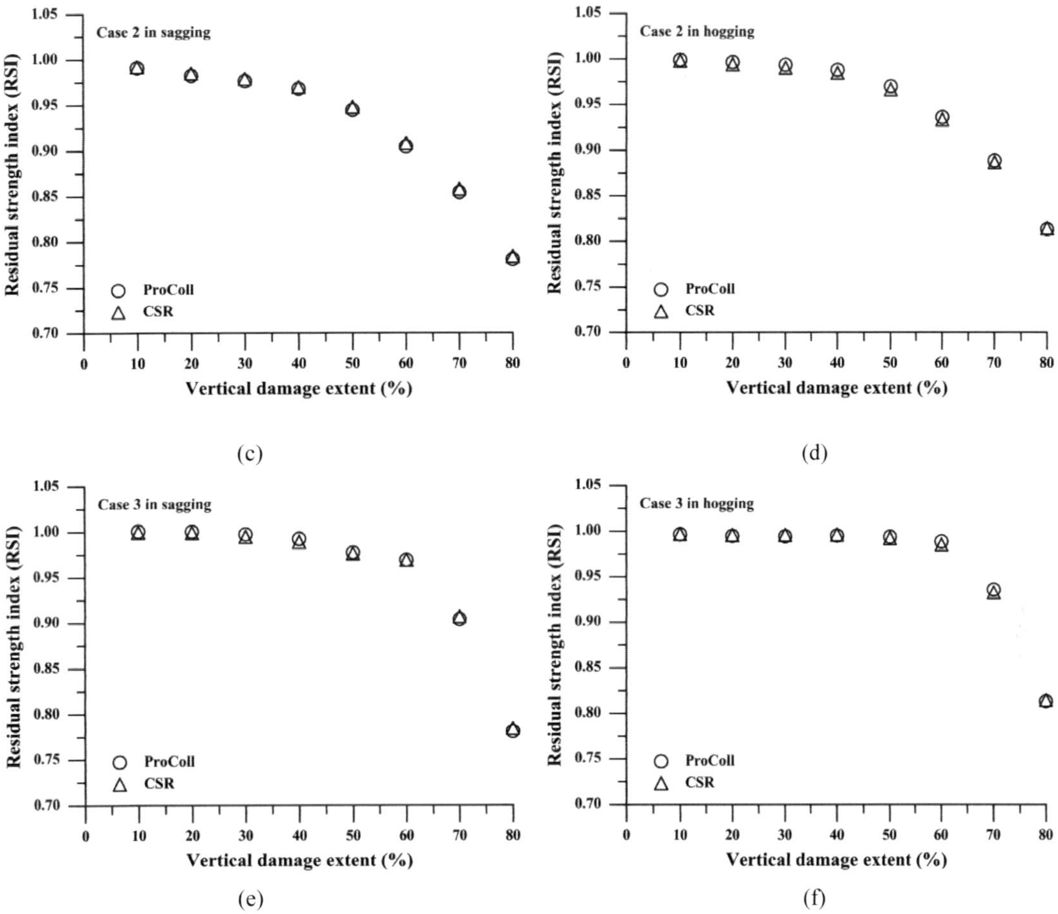

Figure 18. Residual strength index in terms of vertical damage extent

5. CONCLUSIONS

The paper presents an evaluation of the ultimate strength performance of a container ship with side shell damages. A variety of damaged extent and locations are considered. The prediction of the ultimate ship hull girder is completed using a simplified progressive collapse method with two different load-shortening curves formulation. A validation is performed with equivalent nonlinear finite element analysis. The residual safety margin of the container ship is assessed with reference to published hydrostatic and hydrodynamic data and a series of residual strength index diagrams are proposed which may be useful for a rapid prediction of residual strength of container ships with similar class. From this study, the following conclusions may be made:

- A favourable post-collapse response features the sagging collapse of contain ships where a high level of strength still remains;
- The collapse of an intact container ship in hogging experiences a gradual reduction of post-collapse strength. However, it becomes similar to the sagging collapse behaviour if damaged zone is presented;
- The torsion box of the container ship has a dominating influence on the residual ship hull strength. If the torsion box is still intact, the hull girder retains a sufficient strength in comparison to the extreme design load. By contrast, the residual strength may be inadequate if the torsion box has been damaged.
- The ultimate strength performance of damaged container ship appears to be not affected by the rotation of neutral axis

A variety of damage extents and locations have been analysed in the present study. For future work, it is necessary to examine the effect of the longitudinal damage extent on the ultimate ship hull strength. In addition, the consideration of torsional bending and local bottom load may also be important and could introduce further detrimental effect on the ultimate strength performance of the container ships.

6. REFERENCES

1. RS Dow, 'Structural Redundancy and Damage Tolerance in Relation to Ultimate Ship Hull Strength', *In: Advances in Marine Structures*, DERA, Dunfermline, 1997.
2. M Fujikubo, M ZuBair Muis Alie, K Takemura, K Iijima, S Oka. 'Residual Hull Girder Strength of Asymmetrically Damaged Ships', *Journal of the Japan Society of Naval Architects and Ocean Engineering*, 16, 131-140, 2012.
3. S Benson, A AbuBakar, RS Dow. 'A Comparison of Computational Methods to Predict the Progressive Collapse Behaviour of a Damaged Box Girder', *Engineering Structures*, 48, 266-280, 2013.
4. DK Kim, PT Pedersen, JK Paik, XM Zhang, MS Kim. 'Safety Guidelines of Ultimate Hull Girder Strength for Grounded Container Ships', *Safety Science*, 59, 46-54, 2013.
5. M Tekgoz, Y Garbatov, C Guedes Soares. 'Strength Assessment of an Intact and Damaged Container Ship Subjected to Asymmetrical Bending Loadings', *Marine Structures*, 58, 172-198, 2018.
6. PA Frieze, YT Lin. 'Ship Longitudinal Strength Modelling for Reliability Analysis', *In: Proceedings of the Marine Structural Inspection Maintenance and Monitoring Symposium*, SSC/SNAME, Arlington, 1991.
7. JB Caldwell. 'Ultimate Longitudinal Strength', *Trans RINA*, 107, 411-430, 1965.
8. JK Paik, AE Mansour. 'A Simple Formulation for Predicting the Ultimate Strength of Ships', *Journal of Marine Science and Technology*, 1, 52-62, 1995.
9. CS Smith. 'Influence of Local Compressive Failure on Ultimate Longitudinal Strength of a Ship's Hull', *In: Practical Design of Ships and Other Floating Structures*, Tokyo, 73-79, 1977.
10. CS Smith, RS Dow. 'Ultimate Strength of a Ship's Hull under Biaxial Bending, *ARE TR*, 86204, Dunfermline, 1986.
11. Y Ueda, SMH Rashed. 'The idealized structural unit method and its application to deep girder structures', *Computers and Structures*, 18, 277-293, 1984.
12. Y Chen, L Kutt, C Piaszczyk, M Bienek. 'Ultimate strength of ship structures', *Tran. SNAME*, 91, 149-168, 1983.
13. J Amlashi, T Moan. 'Ultimate strength analysis of a bulk carrier hull girder under alternate hold loading condition–A case study Part 1: Nonlinear finite element modelling and ultimate hull girder capacity', *Marine Structures*, 21, 327-352, 2008.
14. Class NK. 'Investigation report on structural safety of large container ships', 2014.
15. S Benson, MD Collette. 'Finite element methods and approaches', *Encyclopedia of Maritime and Offshore Engineering*, 2017.
16. S Benson, J Downes, RS Dow. 'An automated finite element methodology for hull girder progressive collapse analysis', *In: 11th International Marine Design Conference*, Glasgow, UK, 2012.
17. E Alfred Mohammed, HS Chen, SE Hirdaris. 'Global Wave Load Combinations by Cross-Spectral Methods, *Marine Structures*, 29, 131-151, 2012.

7. AUTHORS BIOGRAPHY

Shen Li is a PhD student at Newcastle University. His research mainly focuses on the buckling and ultimate strength of ship structures. Shen Li is an associate member of RINA.

Zhiqiang Hu holds the position of Lloyds Professor of Offshore Engineering at Newcastle University. He is the deputy editor of journal Ocean Engineering and serves as the Advisory Committee member of ITTC, Ocean Space Utilization Committee member of ISSC and Scientific Committee member of ICCGS. Professor Zhiqiang Hu is a member of RINA.

Simon Benson is the academic director of the Marine Propulsion Research Laboratory and Lecturer in Naval Architecture at Newcastle University. He was a member of the ISSC Ultimate Strength Committee. Dr Simon Benson is a member of RINA.

8. APPENDIX

The scantling of the case study 10000TEU container ship is show in Figure A1 with details summarised in Table A1 to A3.

Table A1. Double bottom girders

Girder	Plating	Stiffener
Girder at CL	16.5 AH	150×15 AH FB
Girder 3859 off CL	16.5 AH	200×15 AH FB
Girder 6463 off CL	16.5 AH	240×15/150×15 AH
Girder 8953 off CL	16.5 AH	200×15 AH FB
Girder 11443 off CL	16.5 AH	200×15 AH FB
Girder 13933 off CL	16.5 AH	200×15 AH FB
Girder 17310 off CL	16.5 AH	200×15 AH FB

Table A2. Other stiffeners

Stiffeners	Description
Deck 7 longitudinal	
No. 22-23	300×11 AH BP
No. 24-25	320×12 AH BP
Inner bottom longitudinal	
No. 2-4, 6-7, 10, 12-13, 15-16, 18-19	480×11/200×30 AH
No. 1	480×11/200×15 AH
No. 9	480×11/130×15 AH
No. 20	500×11/200×30 AH
Bottom and bilge longitudinal	
No. 1-4, 6-7, 9-10, 12-13, 15-16, 18-19	500×22/200×30 AH
No. 20	500×25/200×30 AH
No. 22, 30-35	370×13 AH BP

Table A3. Longitudinal bulkheads and side shell longitudinal

Long. No.	L. Bhd. 17310 off CL	L. Bhd. 19082 off CL	Side shell
73		300×75 DH FB	300×75 DH FB
72		400×75 DH FB	400×75 DH FB
71		400×75 DH FB	400×75 DH FB
69		400×75 DH FB	400×75 DH FB
68		400×75 DH FB	400×75 DH FB
67		Deck 2	Deck 2
66		320×11/130×15 AH	400×11/150×20 AH
65		320×11/130×15 AH	400×11/150×20 AH
64		Deck 3	Deck 3
63		400×11/150×20 AH	400×11/200×25 AH
62		400×11/150×20 AH	400×11/200×25 AH
61		400×11/150×20 AH	400×11/200×25 AH
60		400×11/200×20 AH	400×11/200×25 AH
59		400×11/200×20 AH	400×11/200×25 AH
58		Deck 5	Deck 5
57		480×11/200×20 AH	440×11/200×25 AH
56		480×11/200×20 AH	440×11/200×25 AH
55		480×11/200×20 AH	440×11/200×25 AH
54		500×11/200×25 AH	500×11/200×30 AH
53		500×11/200×25 AH	500×11/200×30 AH
52	Deck 7		
51	400×11/200×20 AH		
50	400×11/200×20 AH		
49	240×12 AH BP		

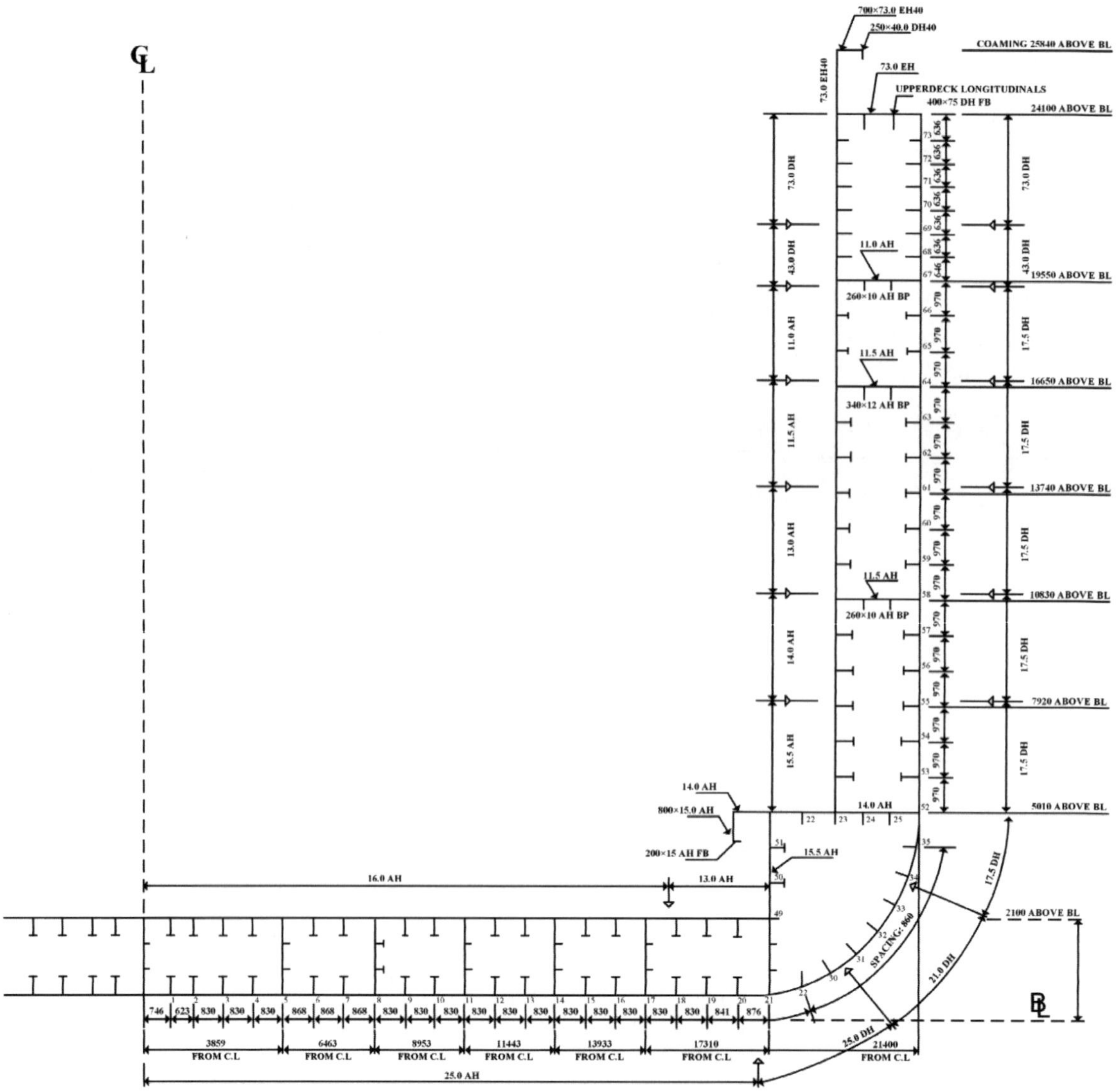

Figure A1. Detailed scantling of the case study 10000TEU container ship

TANKER DAMAGE STABILITY: HISTORICAL PROBLEMS AND CURRENT SOLUTIONS

K. W. Hutchinson, Safinah Group, UK
A. L. Scott, Maritime and Coastguard Agency, UK

SUMMARY

A previous paper read to the Royal Institution of Naval Architects by the authors, at the 2011 'Design and Operation of Tankers' International Conference held in Athens, discussed and illustrated in detail the issues associated at that time with tanker damage stability. It was widely acknowledged by the industry that there was an urgent need to develop guidelines for the verification of damage stability requirements for tankers that regularly sail in conditions of loading significantly different from those in approved documentation, and as a result several international and national guidelines have been developed over the past nine years. This paper summarises the historical problems and then moves on to discuss the solutions developed, namely recently introduced guidelines with particular reference to those published by the United Kingdom's Maritime and Coastguard Agency in mid-2019 i.e. Marine Guidance Note MGN 611 (M).

NOMENCLATURE

B	Breadth (metres)
BHC	IMO Bulk Chemical (Code)
CCKG	Combined (intact and damaged) Critical (fluid) Vertical Centre of Gravity (metres)
CIC	Concentrated Inspection Campaign
CPD	Continuing Professional Development
DSB	Damage Stability Book
EGC	IMO Existing Gas Carrier (Code)
GC	IMO Gas Carrier (Code)
GM_T	(Fluid) Transverse metacentric height (metres)
GZ	Righting Lever (metres)
FPSO	Floating Production Storage and Offloading vessel
FSO	Floating Storage and Offloading vessel
GZ	Righting lever (metres)
GZ_{max}	Maximum positive righting lever (metres) to θ_v
KG_f	Fluid vertical centre of gravity (metres)
KG_s	Solid Vertical Centre of Gravity (metres)
KM_T	Transverse Metacentre above the keel (metres)
IACS	International Association of Classification Societies
IBC	IMO International Bulk Chemical (Code)
ICLL	IMO International Convention on Load Lines
IGC	IMO International Gas Carrier (Code)
IMO	International Maritime Organization
IOPP	IMO MARPOL International Oil Pollution Prevention (Certificate)
ISM	IMO International Safety Management (Code)
LNG	Liquified Natural Gas
LNGC	Liquified Natural Gas Carrier
LPG	Liquified Petroleum Gas
LPGC	Liquified Petroleum Gas Carrier
L_S	Subdivision length (metres)
MCA	UK Maritime and Coastguard Agency
MGN	Marine Guidance Note
MODU	IMO Mobile Offshore Drilling Unit (Code)
MoU	Memorandum of Understanding
MSC	IMO Maritime Safety Committee
MSN	UK Merchant Shipping Notice
PSC	Port State Control
PSCO	Port State Control Officers
Range	Range of positive righting levers measured from θ_e up to θ_v (degrees)
RINA	The Royal Institution of Naval Architects
RO	Recognised Organisation
SDC	IMO Sub-Committee on Ship Design and Construction
SG	Specific Gravity
SIB	Stability Information Booklet
SLF	IMO Sub-Committee on Stability and Load Lines and on Fishing vessels (SDC from 2014)
SOLAS	IMO Safety of Life at Sea Convention
SOLAS90	Version superseding SOLAS60/74 and version prior to SOLAS2009 and applicable to ships constructed prior to 1st January 2009 – the last version based on deterministic damage stability assessment for Passenger ships, Dry Cargo ship damage stability requirements introduced based on a probabilistic assessment
SOLAS2009	Previous version to SOLAS2020 and is applicable to new ships constructed on or after 1st January 2009 – based on a probabilistic damage stability assessment for both Dry Cargo and Passenger ships
SOLAS2020	Current version which is applicable to new ships constructed on or after 1st January 2020 – based on probabilistic damage stability assessment for Dry Cargo and Passenger ships
T&S	Trim and Stability
UK	United Kingdom
UR	IACS Unified Requirement
θ	Angle of Inclination (degrees or radians)
θ_E	Equilibrium angle after asymmetrical flooding in any stage of flooding (degrees)
θ_f	Angle of Downflooding (degrees)
θ_{GZmax}	Angle of Maximum GZ (degrees)
θ_{RoS}	Range of Stability beyond θ_E (degrees)
θ_v	Angle, in any stage of flooding, where the righting lever becomes negative, or the angle at which an opening incapable of being closed weathertight becomes submerged (degrees)

1. INTRODUCTION

The fundamental issue with tanker damage stability lies with the complexity of dealing with fluid loss from a damaged tank which may be filled to any level and carrying liquids with a wide potential range of specific gravity (SG). When this is allied to an extensive variety of loading patterns with empty, part-filled or full tanks and multi-compartment side or bottom / raking damages, demonstration of compliance can be problematic even though the regulations permit the simplifying assumption that all the fluid in the damaged compartment is instantaneously lost and replaced by seawater up to the outside damaged equilibrium waterline. For some tankers, such as crude oil tankers and liquified petroleum and natural gas carriers (LPGCs / LNGCs), the loading pattern, cargo SG and tank filling levels are often predictable and hence damage stability can be verified against pre-approved loading conditions in the approved Trim and Stability (T&S) Book / Stability Information Booklet (SIB).

However, as soon as a loading condition deviates 'significantly' from the approved conditions in the T&S Book / SIB, which is quite common for parcel tankers, products carriers, chemical tankers, Floating Production Storage and Offloading vessels (FPSOs) and Floating Storage and Offloading vessels (FSOs), then some of the regulations require a new damage stability verification which must be undertaken and be approved by the authorities for compliance before the ship sails. It is possible to demonstrate compliance by consulting 'simplified stability information' namely 'condition' or 'generic' combined (intact and damaged) critical limiting KG_f or GM_T curves - however, in order to fully cover the wide range of possible loading scenarios such information is difficult to produce and present in a concise and accessible manner and hence their application is often not 'simple' or quick. The other option is through 'direct' verification of the loading condition using an onboard Stability Instrument or shore-based loading computer - although in practice, for various reasons, advance approval of all the potential damage scenarios for 'non-standard' loading conditions is seldom sought. Over a decade ago, inspections of tankers established, in many instances, that such verification and approval of damage stability compliance was not being undertaken. It was also revealed that there was a considerable lack of awareness of the complexity of damage stability calculations, with many loading officers believing, for example, that compliance with the intact stability criteria alone was sufficient.

The authors discussed and illustrated the above issues associated with the establishment of compliance of tankers with the required damage stability standards in detail almost nine years ago in a paper [1] read to the Royal Institution of Naval Architects (RINA) at their International Conference on the 'Design and Operation of Tankers' held in Athens in June 2011. In the intervening period it was widely acknowledged by the industry that there was an urgent need to develop guidelines for the verification of damage stability requirements for tankers that regularly sail in conditions of loading significantly different from those in approved documentation, and as a result several international and national guidelines have been developed over the past few years.

The following paper briefly summarises the historical problems outlined previously [1], and then moves on to discuss the recently introduced guidelines and particularly the Marine Guidance Note (MGN) published mid last year by the United Kingdom's (UK) Maritime and Coastguard Agency (MCA), namely MGN 611 (M) [2]. This paper will therefore conclude the subject first highlighted by the authors in their paper of nine years ago and also bring under one 'cover' four articles [3 to 6] published in RINA's The Naval Architect journal, by discussing these issues from the aspects of the ship designer, regulators and operators.

2. CURRENT APPLICABLE DAMAGE STABILITY REGULATIONS

The transportation of hydrocarbons of various grades and other liquids by products tankers is covered by the International Maritime Organization's (IMO) [7] International Convention for the Prevention of Pollution from Ships (MARPOL) [8] Annex I 'Regulations for the Prevention of Pollution by Oil'. However, obviously The International Convention for Safety of Life at Sea (SOLAS) [9 to 11] also applies to elements of the design and construction of such ships. The transportation of chemicals in bulk by sea is covered by SOLAS Chapter VII 'Carriage of dangerous goods' and MARPOL [8] Annex II 'Regulations for the Control of Pollution by Noxious Liquid Substances in Bulk'. Both the above Conventions require chemical tankers built after 1st July 1986 to also comply with the International Bulk Chemical (IBC) Code [12], and for chemical tankers constructed before 1st July 1986 Regulation 11 of MARPOL Annex II requires compliance with the Bulk Chemical (BHC) Code [13], which preceded the IBC Code but remains as a recommendation under SOLAS.

The current version of SOLAS, namely SOLAS2020 [9 to 11], covers damage stability for passenger ships of all sizes and dry cargo ships with length greater than or equal to 80 metres with keels laid on or after 1st January 2020 – SOLAS2020 is discussed by the authors in [14] [15]. Certain cargo ship types and specifically tankers are excluded from complying with the probabilistic damage stability regulations in Part B-1 of SOLAS if they comply with the specific other deterministic damage stability requirements included in the following instruments instead:

- **Product and general Oil Tankers** (except combination carriers with Type B freeboards): Regulation 28 of Annex 1 to The International Convention for the Prevention of Pollution from Ships (MARPOL) 73/78 [8];
- **Chemical Tankers**: Chapter 2 of the 2016 International Bulk Chemical (IBC) Code [12] which supersedes the Bulk Chemical (BHC) Code [13] – note that under Chapter 17 cargos are categorised according to their associated environmental and safety hazards namely Type 1 'very severe', Type 2 'appreciably severe' and Type 3 'sufficiently severe', and hence a Type 1 chemical tanker is required to survive the most severe standard of damage and its cargo tanks be located at the maximum prescribed distance inboard from the shell to preclude cargo loss whereas Types 2 and 3 have progressively lesser requirements;
- **Liquefied Gas Carriers**: Chapter 2 of the 2016 International Gas Carrier (IGC) Code [16], which supersedes the Gas Carrier (GC) Code [17] and obviously the Existing Gas Carrier (EGC) Code [18].

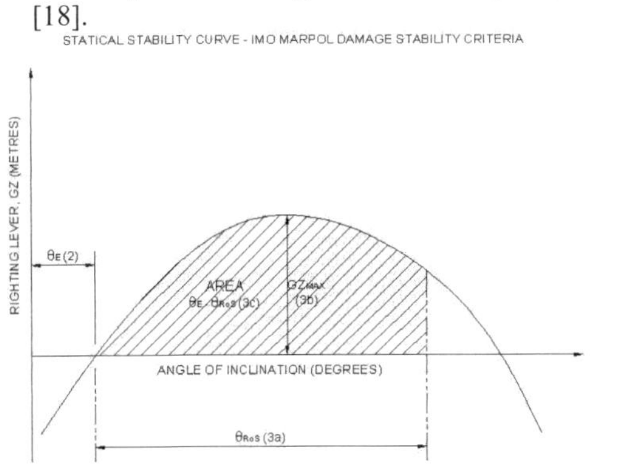

1 The final waterline, taking into account sinkage, heel and trim, shall be below the lower edge of any opening through which progressive flooding may take place. Such openings shall include air-pipes and those which are closed by means of weathertight doors or hatch covers and may exclude those openings closed by means of watertight manhole covers; small watertight cargo tank hatch covers etc.

2 In the final stage of flooding, the angle of heel due to unsymmetrical flooding (θ_E) shall not exceed 25 degrees, provided that this angle may be increased up to 30 degrees if no deck edge immersion occurs.

3a The stability in the final stage of flooding shall be investigated and may be regarded as sufficient if the righting lever (GZ) curve has at least a range of 20 degrees (θ_{ROS}) beyond the position of equilibrium (θ_E).

3b The maximum residual GZ shall be at least 0.1 metres within the 20 degrees range.

3c The area under the GZ curve within the 20 degrees range shall not be less than 0.0175 metre-radians.

3d Unprotected openings shall not be immersed within the 20 degrees range unless the space concerned is assumed flooded. Within this range, the immersion of any openings described in subparagraph 3.1, and other openings capable of being closed watertight may be permitted.

Figure 1: IMO MARPOL Damage Stability Criteria

All tankers / carriers of more than 150 metres in length with Type A freeboards must also comply with the damage stability provisions of Regulation 27 of the International Convention on Load Lines (ICLL) [19] as amended by [20]. This involves meeting specific damage stability criteria following flooding of any (assumed empty) single compartment (not two compartments, typical in most other Codes discussed above) when subject to side damage penetrating to the lesser of B/5 or 11.5 metres with unlimited vertical extent. Should any main transverse bulkheads be spaced less than the lower of 1/3 $L^{2/3}$ or 14.5 metres apart then two compartment damages must be considered.

The damage requirements (extents, permeabilities, criteria etc.) for all the above Codes are summarised together in SLF 52/INF.4 [21] and IACS Recommendation 110 [22].

FPSOs and FSOs etc. are not explicitly covered by MARPOL [8] etc. and ultimately, the applicable regulations are the responsibility of the Flag State / Shelf Regulator. However, it is typical to apply MARPOL damage stability criteria and, in some cases, the damage requirements of Chapter 3 of the 2009 Mobile Offshore Drilling Unit (MODU) Code [23] (see Figure 5 of [1]), either in isolation or in addition to MARPOL, to such vessels.

3. HISTORICAL PROBLEMS

Some of the regulations relating to tanker damage stability state that if a proposed loading condition differs significantly from an equivalent condition in the T&S Book / SIB on board then it must be approved by the authorities for compliance with the damage stability regulations before a tanker sails. Complications arise through having to calculate fluid loss from a damaged compartment which may be filled to any level and carry liquids with a wide potential range of specific gravity. When this is allied to an extensive variety of loading patterns and multi-compartment damages, demonstration of compliance, other than by use of an on-board 'direct calculation' computer, can be difficult even though the regulations permit the simplifying assumption that all the fluid in the damaged compartment is instantaneously lost and replaced by sea-water up to the outside damaged equilibrium waterline. In practice, for various reasons, advance approval of all the potential damage cases for 'non-standard' loading conditions is seldom sought and hence it was widely acknowledged that there was an urgent need to develop guidelines for the verification of damage stability requirements for tankers that regularly sail in conditions of loading that are significantly different from those in approved documentation. Such guidelines have recently been adopted at the IMO and are discussed later in this paper.

Figure 2: Chemical Tanker leaving the River Tyne [Dennis Maccoy, 1989]

3.1 2005 SHIP-OWNERS SURVEY

In 2005 concerns began to arise that some tankers were being loaded only in accordance with the intact stability requirements and that compliance with the damage stability regulations was only being partially considered or, in some cases, overlooked altogether. The specific issues were:

a) low or zero margins on stability in the worst cases of damage;
b) damage cases omitted particularly those of lesser extent;
c) insufficient consideration given to the effects of empty or partially filled tanks;
d) deck tanks either not accounted for or given insufficient consideration; and
e) approvals based upon unrealistic conditions of loading with respect to cargo distribution.

One survey showed that of 11 LPGCs and chemical tankers, with existing stability approvals considered for endorsement at this time, seven (63%) were found to be incorrectly based and omitted critical elements to the verification.

In a more detailed survey covering 76 oil, product and chemical tankers and gas carriers it was found that: 59 (77%) loaded 'non-standard' conditions to those in the approved T&S Book / SIB, including Longitudinal Strength; 43 (56%) regularly operated with slack or empty tanks; 69 (90%) were fitted with Loading Instruments, however 30 (39%) could only assess intact stability and 16 (21%) could only assess 'non-standard' loading conditions using an intact stability Loading Instrument.

Additionally, from Port State Control (PSC) inspections, it was observed that tankers were regularly loaded in 'non-standard' conditions from those in the approved T&S Book / SIB. In addition, it was found that tanker's loading conditions were commonly being appraised using a longitudinal strength computer with only intact stability assessment, and that Masters were utilising loading conditions clearly marked 'not valid for damage'.

The conclusions from the various surveys and inspections undertaken during this period were that insufficient pre-departure checks to ensure compliance with statutory damage stability requirements were being made and that existing stability approvals for some tankers may not be sufficient to ensure that minimum statutory requirements are being met; it was also concluded that there was a compelling need for further action to ensure existing international instruments are being complied with.

The results of this work were informally presented to IMO at SLF 51 (Stability, Load Lines and Fishing Vessel Safety Sub-Committee) in 2009.

3.2 2009 DATA GATHERING EXERCISE

As a result of various 'flag-in' issues and the 2005 survey and PSC inspections, during 2009 one IMO Member State conducted a confidential survey of Masters and Officers of tankers to gather information on the facilities available to them to reliably check the stability (intact and damaged) of a ship, and that ships were loaded in accordance with the approved damage stability information / regulations.

For this questionnaires and checklists were used, as detailed in [1], to gather information regarding the application of damage stability to tankers in support of on-going work at IMO aimed at:

a) ensuring existing IMO instruments were enforced in a consistent manner;
b) minimising risks to tankers and the seafarers who operate them;
c) minimising risks to the environment from loss of marine pollutants; and
d) ensuring non-compliant tankers are identified and brought into compliance.

The survey of 73 tankers was completed by the end of September 2009 with a mixture of oil, chemical, chemical / products tankers and gas carriers. It was found that, 47 (64%) were either fitted with satisfactory means of routinely verifying that the damage stability requirements were met on departure, or that there was no requirement for such verification due to use of standard loading conditions. It was also established that, when inspected, 23 (31%) were loaded to a condition significantly different from an approved loading condition and had no information available to verify compliance with damage stability or that the information available was not routinely used - hence, these tankers were considered non-compliant.

The data from this together with reanalysed data from the 2005 Owners Survey was presented to SLF 52 in January 2010 [1] showing that 35% of ships were non-compliant.

3.3 2010 PARIS MOU CIC

In 2010, the Paris Memorandum of Understanding (MoU) on PSC undertook a three-month Concentrated Inspection Campaign (CIC) on Tanker Damage Stability, between 1st September 2010 and 30th November 2010, to verify correct damage stability on oil tankers, chemical tankers and gas carriers, as discussed in [1].

During the CIC, the Paris MoU Maritime Authorities conducted a total of 1,065 inspections - 419 oil tankers, 538 chemical tankers and 108 gas carriers. A total of 94 (8.8%) inspections resulted in deficiencies directly related to the CIC leading to 4 (2 oil tankers and 2 chemical tankers) being detained for not complying with damage stability requirements. Significantly, for a total of 173 ships (16.2%) comprising 77 oil tankers, 84 chemical tankers and 12 gas carriers, the Master could not demonstrate that they were normally loaded in accordance with the approved T&S Book / SIB.

Detailed analysis of the results of the campaign were reviewed by the 44th meeting of the Paris MoU PSC Committee in May 2011, and a report subsequently submitted to IMO.

3.4 2010 BLACK SEA MOU CIC

The Black Sea MoU undertook their CIC in conjunction with routine PSC inspections. To harmonise with the Paris MoU, the same questionnaire and checklist was utilised over the same period.

The CIC conducted 229 inspections on board 227 tankers, which equated to 95% of all tankers entering ports of Black Sea MoU during the period. Deficiencies were found on about 41% of the inspected tankers but none led to detention. It was found that 97% of the inspected tankers had an approved T&S Book / SIB on-board and that for 96% of the time it was written in the language understandable by the Master. For 89% of inspected tankers the Master could demonstrate normal loading of the ship with the T&S Book / SIB.

Detailed analysis of the results of the campaign were considered by the Black Sea MoU PSC Committee in March 2011 and, as with those from the Paris MoU, were submitted to IMO.

3.5 ISSUES WITH TANKERS

The fundamental difficulties in demonstrating that tankers comply with damage stability arise from having to address, often simultaneously:

a) fluid loss from a damaged tank;
b) an extensive variety of loading patterns, with empty, part-filled or full tanks sometimes combined with a range of cargo SGs; and
c) multi-compartment side / bottom / raking damages.

3.5 (a) Damage Scenarios

Damage scenarios required by MARPOL [8], the IBC [12] and IGC [16] Codes are numerous. Potential multi-compartment side, bottom and raking damage cases required in the regulations are, as discussed and illustrated in Figures 6 to 19 of [1], very complex and numerous in nature, possibly totalling many hundreds of cases. Hence, the identification of the dominant case(s) from amongst the typically large operational draught range of all tankers (which are often 'pure' or 'semi' deadweight carriers) is obviously complex and time consuming, and therefore difficult to establish and present in an accessible form.

3.5 (b) Fluid Loss and Cargo Specific Gravity

As alluded to above, a major issue with tanker damage stability lies with the complexity of dealing with fluid loss, so-called fluid deadweight 'drop-out', from damaged tanks which may be filled to any level and carrying liquids with a wide potential range of SGs. For example ships, such as some products and chemical tankers, are designed to carry a range of cargos, with SGs ranging from 0.7 to 2.1 tonnes.metres^{-3}, in any given loading condition.

As an illustration, if the SG of the liquid in a damaged tank, initially assumed to be full, exceeds that of seawater then the ship could list away from the damaged side in certain circumstances. Conversely, if the SG of the liquid in a damaged cargo tank, initially assumed full, is less than that of seawater then the ship could list towards the damaged side in certain circumstances. Thus, accounting for full or partially filled ballast tanks and different extents of damage, as discussed above, it can be appreciated that establishing damage stability compliance involves many permutations and hence is complex and time consuming. This is illustrated graphically in Figures 20 to 26 of [1].

3.5 (c) Loading Patterns

When different loading patterns are allied to an extensive variety of damage cases the demonstration of compliance with the damage stability criteria becomes difficult.

For some ships, such as crude oil tankers as illustrated in Figures 27 to 32 of [1], the loading pattern, cargo SG and tank filling levels are often predictable. Hence, damage stability performance can be verified with some confidence against pre-approved loading conditions detailed at build in the T&S Book / SIB, with the detailed assessment given in the Damage Stability Book (DSB).

However, as soon as a loading condition deviates 'significantly' [22] [24] [25] from the fixed approved loading conditions, which is quite common for parcel tankers, products carriers, chemical tankers (often designed to carry a range of cargos with various SGs), FSOs and FPSOs, then, strictly speaking, a new damage stability verification calculation is needed which should be approved by the certifying authority before the ship sails / prior to loading offshore. Examples of such 'non-standard' loading are given in Figures 36 to 42 of [1].

Hence, when such different 'non-standard' loading patterns, consisting of combinations of various tank filling depths, possibly with different cargo SGs, are allied to an extensive variety of damage scenarios required by the tanker regulations it is obvious that demonstrating damage compliance is all but impossible other than by use of a Loading Instrument. This is due to the complexity and subsequent volume of information required to adequately convey the post-damage performance of the ship to the Master.

However, for some tanker types, such as crude oil tankers etc. mentioned above, which typically one cargo SG, then it is possible that intact and complex damaged critical limiting maximum Fluid Vertical Centre of Gravity (KG_f) or minimum Fluid Transverse Metacentric Height (GM_T) curves, such as Combined Critical KG_f ($CCKG$) curves as shown in Figures 3 to 5 and discussed below, could be constructed and provided but, as discussed, they are not 'simple' or quick to use by the crew.

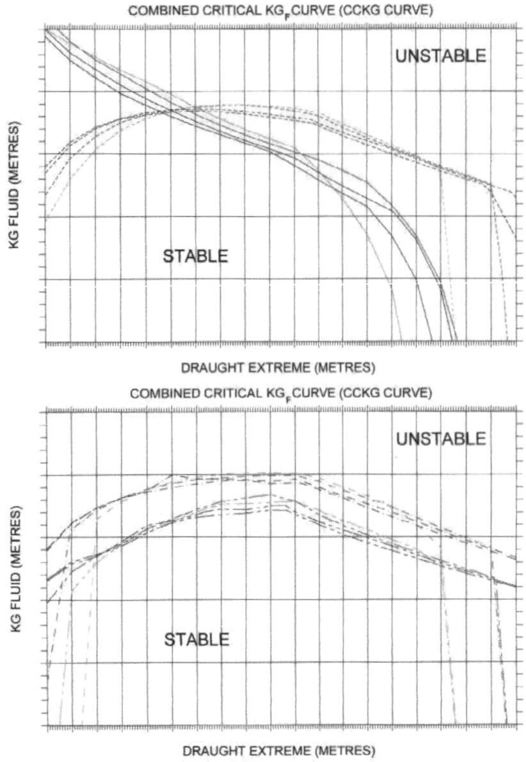

Figure 3: Typical intact and side damage critical envelope KG_f curves, various trims

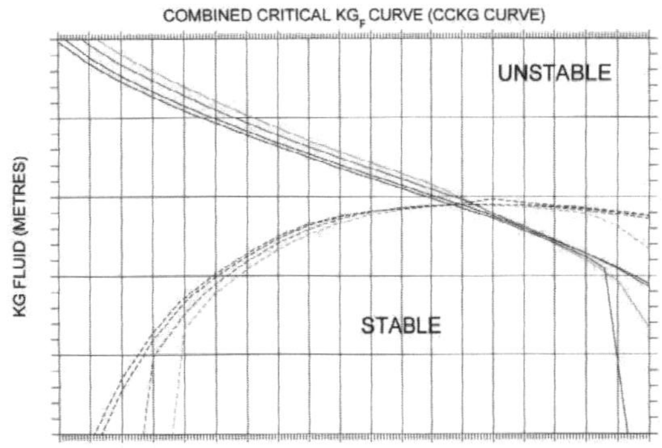

Figure 4: Typical intact and raking damage critical envelope KG_f curves, various trims

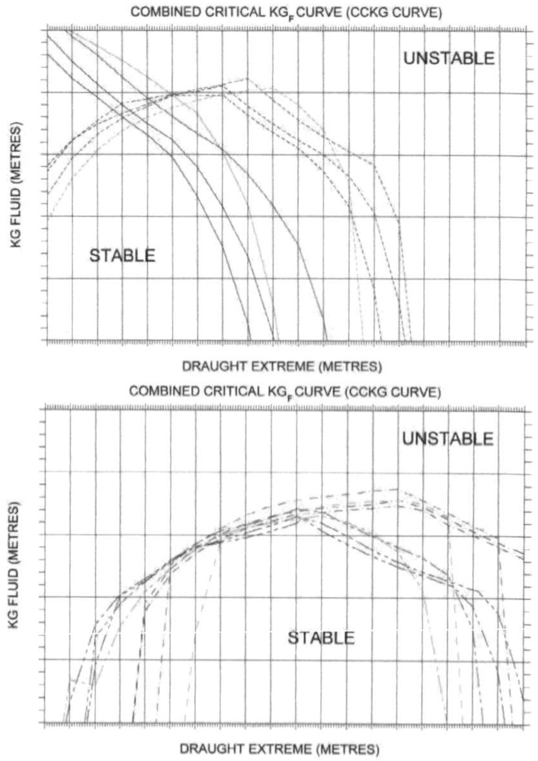

Figure 5: Typical intact and bottom damage critical envelope KG_f curves, various trims

3.5 (d) Variation in Damage Stability Performance

It has been shown that, within the seafaring community, there was a significant lack of awareness of the complexity of damage stability calculations together with their variation due to loading – demonstrated by Figure 6. This is surely an issue that must be addressed through ongoing education (Continuing Professional Development, CPD) of Naval Architects and, most importantly, seafarers. Additionally, such insight needs to be incorporated into the education of all new Naval Architects and deck cadets / officers.

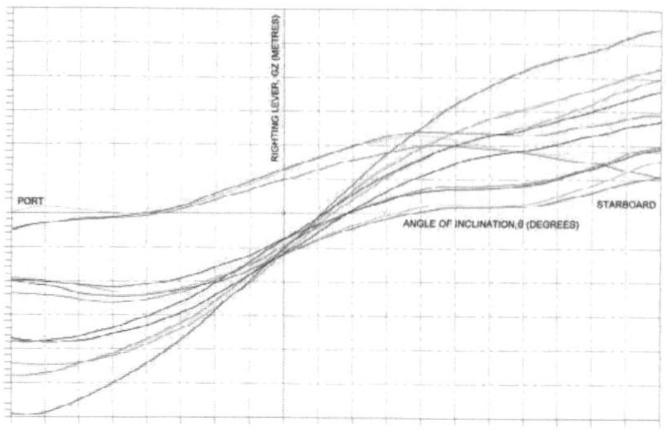

Figure 6: Righting Lever (GZ) Curves for one Damage Case but various Loading Conditions

3.5 (e) Presentation and establishing compliance

As stated above, it is possible, but exceptional in tanker T&S Books / SIBs, to demonstrate damage compliance of 'non-standard' loading conditions, against an extensive variety of damage cases using *CCKG* information. This is 'simplified stability information' namely 'loading condition specific' or 'generic' combined (intact and damaged) critical limiting maximum KG_f or minimum GM_T curves, as illustrated in Figures 33 to 35 of [1], and tables. Such information is common in the T&S Books / SIBs of SOLAS ships (both using the previous SOLAS90 [26] deterministic and probabilistic approach to damage applied in SOLAS2009 [9] and the current SOLAS2020 regulations [9 to 11] as discussed by the authors in [14] [15] and also [27] [28]) as the stability analyses typically assumes a 'bare boat' basis where it is assumed that there is no fluid deadweight 'drop-out' as there is relatively little fluid available for loss compared to the buoyancy or internal volume of the ship. However, to fully cover the wide range of possible loading condition scenarios and evolutions and accurately assess the damage performance of tankers fluid deadweight 'drop-out' must be considered. Such *CCKG* information is not easy to produce and present concisely and yet this is of the utmost importance from the aspect of effective and correct use by the Master or crew to ensure the operational safety of tankers, their crews and the environment. The production of such *CCKG* information is difficult for the Naval Architect which is perhaps why most tanker shipbuilders do not provide *CCKG* information for the use of the Master and, if it is provided, application on-board is neither 'simple' or quick.

Figure 7 shows a group of critical limiting maximum KG_f curves for a tanker for one assumed degree of filling and SG – which by coincidence is 'bare boat'. The figure is complex and congested as it contains both intact and damage critical limiting curves covering the operational trim range of the ship and all cases of side, bottom and raking damage together with 'exception' cases required to allow the developed loading conditions to be proved compliant against the combined (intact and damage) critical limiting maximum KG_f curves, rather than just by 'direct calculation'. The identification of such exception cases is a complex and time-consuming process and highly dependent upon the range and extent of loading conditions developed. Other similar groups of curves would be required to adequately cover the fill range of the cargo oil and water ballast tanks, namely: nine for 0% / 50% / maximum fill steps; and, twenty-five for 0% / 25% / 50% / 75% / maximum fill steps; etc.

Obviously, the use of such curves is impracticable and hence, as with intact only curves, it is usual to deconstruct them and group them in 'damage specific' combined critical limiting maximum KG_f curves (see Figure 8). However, splitting the curves into such

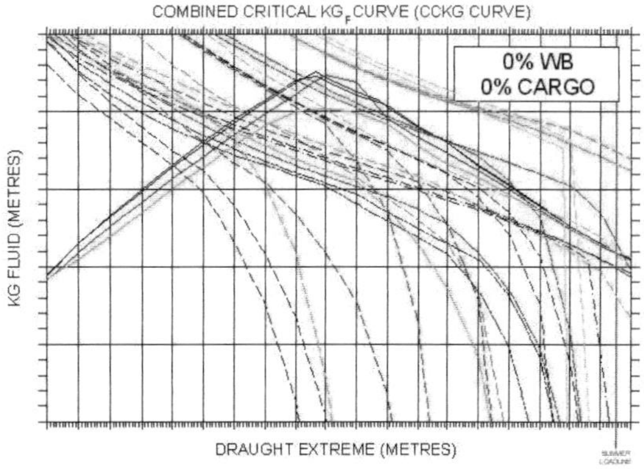

Figure 7: Typical intact and damage critical envelope KG_f curves, various trims (zone exceptions)

'groups' can result in up to seventy-five separate curves for a typical large oil tanker, which would be increased again if consumables such as diesel and fresh water etc. are also considered. The curves described in Figure 8 are 'generic' in nature as the fill levels in each zone can be checked individually and are hence independent of the loading conditions, which is one way that 'deadweight drop-out' curves are commonly produced. However, they are still dependent upon the SG of the cargo and if this varies then 'SG specific' curves would also have to be provided.

It is more common in tankers, if combined critical limiting maximum KG_f curves are presented, to do so in the form of 'condition specific' curves such as this shown in Figure 9. These curves present specific sets of curves for 'typical' loading conditions and advise any specific requirements for loading, of either cargo or ballast tanks, to comply with damage stability - also shown on these are curves for differing SGs.

It should be noted that Figure 9 does not account for differing levels of consumables such, such as fresh water and oils etc., but only cargo and water ballast. However, for some designs the level of these can be crucial regarding damage stability and the production of *CCKG* curves that accurately describe the maximum operational envelope of the ship.

Hence, the application of any *CCKG* curves obviously requires substantial time for interrogation, interpolation and calculation to ascertain if the ship is safely loaded. Therefore, many shipbuilders only give intact critical limiting maximum KG_f curves and only provide 'direct calculations' (in the Damage Stability Book, DSB) on the loading conditions given in the SIB which limits the loading of the ship to these conditions and promulgates the misconception among loading officers and Masters.

The alternative, preferred - as discussed later in this paper, option is through 'direct' verification of the condition using an on-board or shore-based loading computer, but the interpretation of information produced

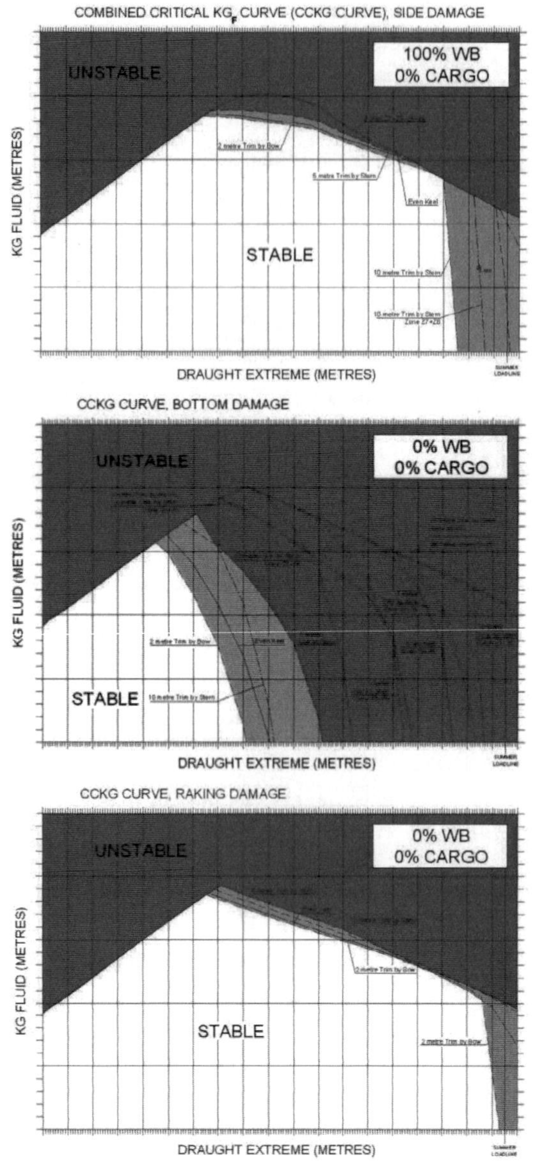

Figure 8: Typical 'generic' *CCKG* curves

is also a potential issue. Additionally, if the on-board Loading Instrument is in error or fails then if *CCKG* information is not available to the Master within the T&S Book / SIB (which is the primary source of stability information for any ship rather than the Loading Instrument) then the ship cannot proceed to sea or load accordingly if at sea. As stated above, it is uncommon for shipbuilders, consultants etc. to present *CCKG* information in tanker T&S Book / SIB to allow the Master to demonstrate compliance with damage stability. It is more common, but still fairly rare, for *CCKG* information to be derived and presented for FPSOs and FSOs, especially conversions, where the stability analysis and associated Statutory documentation is produced by consultants. It was not unusual for tankers to be provided with critical KG_f information covering compliance with the intact stability criteria only and it was clearly found, during the above-mentioned surveys, that some loading officers believed that compliance with the intact stability criteria alone was sufficient.

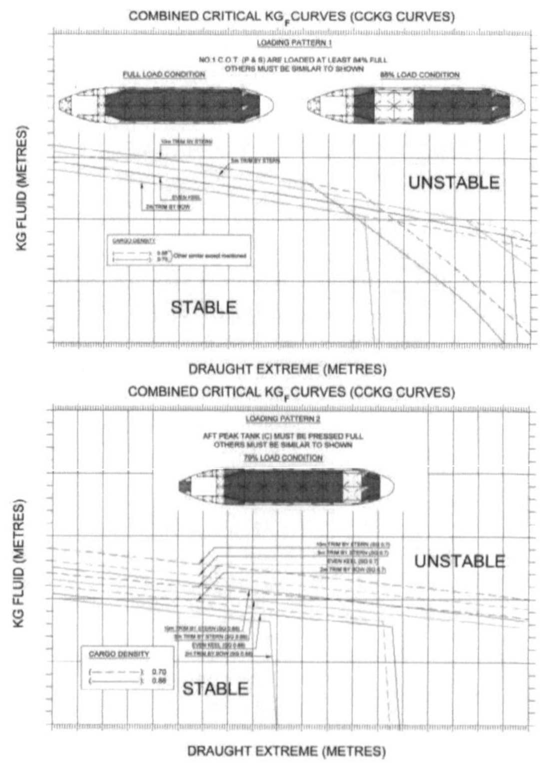

Figure 9: Typical 'condition specific' *CCKG* curves (with various fills and SG)

From a purely professional perspective, it is perhaps surprising that Naval Architects involved with the design, conversion or provision of operational support for such 'liquid tank deadweight ships' in some cases seemed to overlook the fact that the damage stability critical KG_f curves usually dominate over those derived from intact stability, especially at deeper draughts. The authors found this rather alarming given that, a decade or so ago, both owners / operators and regulators within the tanker industry were made aware of a considerable lack of awareness amongst all parties of the complexity of damage stability calculations including the inherent susceptibility of a tanker's damage stability performance to varying cargo types, tank filling depths, cargo SGs etc. It is imperative that the Master is furnished with this understanding in a clear, unambiguous and accessible manner together with any associated restrictions such as 'liquid loading limits' etc.

As discussed in detail in Section 4 directly following this, the earlier provision of such misleading information and its approval by the authorities is being addressed following recent revisions to MARPOL [8] and related tanker Codes [12] [16] which, from 2016, require the use of on-board or shore-based computers to supplement the approved T&S Book / SIB for all new tankers with a phase-in period of five years until 2021 for existing tankers (with only limited exceptions). For ships other than tankers, IMO is making it clear in the forthcoming amendments to SOLAS that *CCKG* data must combine both the intact and damage stability criteria in one consolidated curve or set of curves.

4. SOLUTIONS TO THE HISTORICAL PROBLEMS

In this Section, the solutions proposed and adopted to demonstrating a tanker's compliance with the damage stability regulations including the provision of Guidelines by the IMO are explored.

Noting some of the long-standing technical problems associated with demonstrating a tanker's compliance with the damage stability regulations in MARPOL [8], the IBC Code [12], the BCH Code [13], the IGC Code [16], the GC Code [17] and the EGC Code [18], on 22nd May 2014 MARPOL Annex 1 was amended by the addition of Chapter 4, Paragraph 28(6) and Chapter 1, Paragraph 3(6) which states that all oil tankers constructed on or after 1st January 2016 shall be fitted with a stability Loading Instrument capable of verifying compliance with not only intact but also damage stability requirements. Oil tankers constructed before 1st January 2016 shall be so fitted with such a Loading Instrument at the first scheduled renewal survey of the ship on or after 1st January 2016 but not later than 1st January 2021, although this requirement may be waived if it can be shown that the existing system for verifying damage stability is satisfactory.

Similar amendments were made to ships constructed under the remaining Codes listed above, except that for the IGC Code [16] where the amendments apply to ships constructed prior to 1st July 2016 with compliance required at the first renewal survey after this date and no later than 1st July 2021. In addition, the International Oil Pollution Prevention (IOPP) Certificate in MARPOL Appendix II (Form B) and the Certificates of Fitness in the remaining Codes were amended with statements to the effect that either the tanker has been fitted with an approved stability Loading Instrument in accordance with Regulation 28(6) (or equivalent in the remaining Codes) or that the fitting has been waived in accordance with Regulation 3(6) (or equivalent).

4.1 THE IMO GUIDELINES ON TANKER DAMAGE STABILITY COMPUTERS

From the above amendments to MARPOL [8] etc. it can be seen that it was concluded at IMO that the solution to the technical problems of demonstrating compliance with the damage stability regulations, as outlined previously in Section 3, is to employ suitable damage stability computers either on board ship (Loading Instrument) or at a shore station with suitable communication links. Recognising the need for a fuller explanation and more background information, IMO produced circular MSC.1/Circ.1461 entitled *Guidelines for Verification of Damage Stability Requirements for Tankers* [29] which was approved at the 92nd Session of IMO's Maritime Safety Committee (MSC 92) on 8th July 2013. The Guidelines are in two parts:

Part 1. Defines how approval of damage stability calculations or stability Loading Instruments for new oil tankers, chemical tankers and gas carriers constructed on or after 14th June 2013 should be conducted. Although it directly applies to new approvals, Part 1 summarises all pre-existing guidance with which any previous damage stability approvals should have originally been made, hence it may also be used to assess the validity of an existing approval. Technically, an existing stability approval which does not follow the Guidelines is deficient as it does not follow the original Instruments against which it is approved e.g. MARPOL Annex 1. It includes details of the qualifications of the personnel involved, the plans and data to be supplied, the calculation procedures and advice regarding the need to consider all relevant damage cases, including lesser cases of damage to both sides and bottom of the ship. It also describes the permissible limits within which damage stability calculations or a damage stability Loading Instrument must lie when checked for accuracy.

Part 2. Is mainly intended for the guidance of third parties, such as ship operators, ship's officers and Port State Control Officers (PSCO). Section 6 describes the records required to be kept on board all tankers and gas carriers in order to demonstrate damage stability compliance for the approved methods of verification listed in Section 4. The processes and records described here are also applicable to existing ships and Section 6 should be considered mandatory once re-certification of existing tankers is completed under the revised Instruments. Part 2 provides guidance for operators and ship's officers to meet their obligations under the ISM Code, and to third parties such as PSCO and International Safety Management (ISM) auditors.

To ensure that tankers can have their damage stability reliably verified on board they must either be fitted with an approved stability Loading Instrument (the default position for new ships unless there are compelling reasons not to comply) or, for existing tankers, certified to continue to use a validated and approved existing method. All existing damage stability approvals in place

Figure 10: Oil Tanker

Figure 11: Chemical Tanker

at the time the Guidelines came into force remain valid provided they meet the standards from Part 1 (as this repeats the required practice at the time of the original approval) and on condition that any guidance is being correctly followed.

The Guidelines were developed by the IMO well before the applicable enforcement date for the changes to MARPOL [8] and the other Codes, namely 1st January 2016. This was to allow time for operators to assess what changes, if any, were required to the stability verification methods used on their existing tankers and, if deemed necessary, to order new computer systems (on-board Loading Instruments) and have them approved in time to comply with the new regulations within the phase-in period.

Following the phase-in of changes to the applicable Instruments and re-certification, the default method of damage stability verification shall be carriage of a Type 3 stability Loading Instrument, which is discussed below, capable of verification by direct calculation. Other acceptable methods of verification, if retained from before re-certification, must be validated against the Guidelines and then authorised under a waiver on the IOPP Form B or Certificate of Fitness. These accepted methods include operating the tanker closely in accordance with, but not outside the range of, the loading conditions given in the approved T&S Book / SIB.

4.2 TYPE 3 TANKER DAMAGE STABILITY COMPUTERS

The International Association of Classification Societies (IACS) had previously introduced a Unified Requirement (UR) relating to stability computation under UR L5 [25], with the objective that any Loading Instrument fitted on a ship classed with an IACS member contracted after 1st July 2005, which incorporates a stability element, should be approved for verification of all stability requirements that apply, including damage. Under these requirements, a Loading Instrument utilising Type 3 software is capable of automatically calculating both intact and damage stability by direct application of pre-programmed damage cases for each loading condition. A clear warning of non-compliance shall be given on screen and in hard copy printout if any of the loading limitations are not complied with. 'Direct calculation' means that the stability Loading Instrument utilises a full 3-D model of the ship including the hull and all buoyant appendages, all damageable compartments / tanks / spaces downflooding points etc., and is programmed to analyse all the potential damage scenarios required by the relevant Code or Instrument.

5. UK MCA PROPOSED SOLUTIONS

This final Section of the paper introduces the UK's MCA's new guidelines for verification of tanker's damage stability compliance by discussing the recently published MGN 611 (M) *Damage Stability: Alternative verification method for tankers – UK interpretations and procedures* [2].

5.1 REQUIREMENT FOR MCA GUIDELINES

Although publication of MSC.1/Circ.1461 by the IMO in 2013 [29] proved immensely useful in clarifying and interpreting the amendments to the various tanker codes [8] [12] [13] [16] [17] [18] through requiring the use of Stability Instruments to verify the damage stability compliance of tankers prior to departure, the MCA found that there remained some areas where further guidance on the UK's policies was needed. It was therefore decided to produce a new MGN to endeavour to clarify UK's policies towards implementation of the amendments to the tanker codes and application of the associated guidelines [29]. It was felt that the new MGN should be addressed to all those involved with the tanker business, from owners and operators to shipbuilders, consultants and software designers as well as those involved with system approval and regular inspection, such as surveyors and PSCOs. It was also felt that the MGN should be comprehensive therefore including as much of the relevant background information as possible in one document to minimise the need for cross-referencing.

Figure 12: Liquified Natural Gas Carrier

Figure 13: Liquified Petroleum Gas Carrier

5.2 DEVELOPMENT OF MCA GUIDELINES

The development of MGN 611 (M) [2] was instigated by the MCA in early 2016 at the request of several Classification Societies and Stability Instrument producers who desired guidance on how to apply IMO Circular MSC.1/Circ.1461 [29]. IMO requires 'new' tankers, constructed on or after 1st January 2016 to be fitted with approved Stability Instruments capable of verifying compliance with the appropriate, both intact and damage, stability regulations. Whereas for 'existing' tankers, constructed before 1st January 2016, some of which may have already been fitted with a Stability Instrument, the Administration must be satisfied that procedures are in place for demonstrating that the stability regulations were being fully complied with prior to departure. If so satisfied, the Administration or its appointed 'Recognised Organisation' (RO), typically a Classification Society, would issue a waiver from the requirement to fit a newly approved Stability Instrument. Existing MARPOL [8], IBC Code [12], BCH Code [13] or EGC Code [18] tankers must comply with these requirements at the first scheduled renewal survey on or after 1st January 2016 but not later than 1st January 2021. For GC Code [17] tankers the compliance dates are the same except that compliance is required at the first periodical (rather than renewal) survey after 1st January 2016, and for IGC Code [16] tankers the final compliance date is 1st July 2021.

Given the above due dates of compliance for existing tankers there was, and remains, a degree of urgency as ships come up for their renewal or periodical surveys. Extensive meetings and discussions were therefore held with representatives from the ROs and the industry and a list of the issues which were causing difficulties was drawn up, and the MCA's policies towards each of which form the basis for the MGN.

5.3 DESCRIPTION OF MCA GUIDELINES

MGN 611 (M) [2] is divided into three parts. The main part commences with an introduction followed by some background information on IMO Circular MSC.1/Circ.1461 [29]. This is followed by sections on the actions to be taken by all the various stakeholders in the tanker business involved with verifying that the ships meet the relevant intact and damage stability requirements. At the end of the main part is a summary and some conclusions.

There then follow two annexes:

- Annex A consists of extracts from MARPOL [8] and the other tanker codes [12] [13] [16] [17] [18] relating to the obligation to fit approved Stability Instruments or demonstrate some other satisfactory means of verifying compliance with the relevant, intact and damage, stability requirements. As alluded to previously, the aim here is to keep the relevant amendments to the damage stability codes together within one document for easy reference;
- Annex B consists of eight Sections:
    - Section 1 gives an introduction and more detailed background to the IMO guidelines in Circular MSC.1/Circ.1461 [29];
    - Section 2 describes the characteristics of the various types of software which can be used to demonstrate compliance with the stability requirements, based on the IACS UR L5 [25] published in June 2017;
    - Section 3 consists of a flow chart to assist stakeholders in the decision-making process as to whether an 'existing' tanker complies with the amended regulations and guidelines for fitting a Stability Instrument. It shows the three possible outcomes, namely: (1) Issuing a waiver from fitting a new approved Stability Instrument; (2) Approving an existing Stability Instrument; or (3) Fitting a new Stability Instrument and having it approved. In the latter case the Stability Instrument should utilise IACS UR L5 Type 3 software - which is also the MCA's strongly recommended preference for use on 'new' tankers constructed on or after 1st January 2016;

- Section 4 expands on the flow chart and sets out in some detail the various options available for issuing waivers, which are ultimately issued by the MCA in liaison with the appointed RO;
- Section 5 is designed to assist front line surveyors and PSCOs in deciding whether a tanker is fully compliant with all aspects of the stability regulations, both intact and damage, and outlines several checks which can be undertaken during surveys or inspections. A spreadsheet is included which summarises the documentary evidence needed to confirm the validity of the verification method being employed, including those ships where stability at departure is verified at shore-based stations rather than on board;
- Section 6 explains how all the above should be integrated into an approved and certified ISM procedure for assessing risks, establishing appropriate safeguards and dealing with the issue of demonstrating compliance with the stability regulations, however that is achieved, to the satisfaction of the MCA / RO and PSCOs. This includes the important issue of ensuring that personnel are suitably qualified and trained to use the Stability Instrument and that there are always at least two such personnel available either on board or ashore whenever needed;
- It was decided to include a short section, Section 7, on the historical background to computing damage stability for tankers, explaining how the increased computational power of modern computers together with the associated enhanced ability of software has facilitated the possibility of performing complex 'direct' damage stability calculations within a reasonable time-scale once the actual loading conditions (departure, mid-voyage and arrival etc.) are known. Existing ships may have software systems on board which do not fully conform to the latest requirements and so decisions may have to be taken as to whether such systems are permissible or whether they must be upgraded;
- Finally, Section 8 attempts to address some of the many issues and questions which have arisen during the application of the IMO guidelines in Circular MSC.1/Circ.1461 [29], fuller details of which are included in the following section of this article.

5.4 APPLICATION OF MCA GUIDELINES

Initial drafts [30] of MGN 611 (M) included a question and answer section in which the MCA attempted to resolve some of the difficult issues raised by those involved with applying the IMO guidelines in Circular MSC.1/Circ.1461 [29]. For the published version of MGN 611 (M) [2], it was decided to integrate these questions and answers into the main body of the text where appropriate; many of them are now included in Section 8 of Annex B. They are too numerous to include fully in this article but below are a couple of examples of the more difficult questions received during the consultation period and how the MCA is proposing they be dealt with.

Question 1: What if the approved damage stability information (such as the DSB) or the SIB (also termed the T&S Book) produces results which are incompatible with those produced by the newly approved Stability Instrument? Does the stability information need to be revised and re-approved so that the results are fully compatible?

MCA Response 1: If the differences are attributable to modern and more accurate calculation methods being used in the approved Stability Instrument as opposed to, for example, the use of 'look-up' tables in the approved Damage Stability (calculation) Book and / or the SIB, then the Stability Instrument may be approved and the SIB will not require to be revised and re-approved. If, however, the differences are due to fundamental changes to the stability information for the ship (including, but not limited to, items such as down-flooding openings, arrangement, maximum draught etc.) then the original damage calculations and SIB must be revised and re-approved in conjunction with the approval of the Stability Instrument.

In general, the SIB includes loading conditions covering the intact situation which are aligned with those in the approved Stability Instrument. If the damage stability calculations have not been performed in accordance with IMO guidelines in Circular MSC.1/Circ.1461 [29] then the SIB is invalidated for evaluating any condition of loading and cannot be used. <u>The results from the approved Stability Instrument take precedence</u>. The SIB should be retained on board and continue to be used for everything apart from validating the damage stability. This must be made clear to the owners, operators and PSCOs by endorsing the SIB accordingly and updating the Document of Compliance for the ISM Code.

Question 2: Is intermediate stage (IS) flooding to be analysed? If so, which of several methods should be used and for how many stages? Is there a danger that calculation time will be excessive, especially if changes are needed to the initial loading condition, resulting in possible delays to the departure of the tanker? Should new Stability Instruments being installed on existing tankers necessarily use the same analysis method for IS flooding as was used in the previously approved damage stability calculations.

MCA Response 1: Section 9 of IMO's Circular MSC.1/Circ.1461 [29] makes clear that IS flooding is to be analysed by the approved Stability Instrument. It is tempting to argue that the final stage of equilibrium is always the worst in terms of residual stability, but this is not always the case. It is therefore necessary for the Stability Instrument to have the capability of calculating IS flooding for five stages by any of the methods indicated in Section 9.3 of MSC.1/Circ.1461. However, the MCA is aware of the risk of over-burdening the Stability Instrument and so the MCA's position is that most of the IS analysis work should be undertaken at the design stage when it ought to be possible to identify and select those damage cases most likely to result in the ship being non-compliant with the regulations during IS flooding. Only these damage cases would then be analysed for IS flooding by the Stability Instrument. This selection of damage cases would occur in conjunction with the Administration or appointed RO during the approval process. A similar selection process should be applied to minor damages (see Annex B, Section 8.3.5 of MGN 611 (M) [2]) to reduce the number of damage cases and hence the computing time whilst ensuring that all damage cases which could result in non-compliance are automatically covered.

It is not essential that new Stability Instruments being installed on existing tankers follow the same procedure for undertaking IS analysis as was used in the existing approved documentation. Just as for Question 1, the results from the new approved Stability Instrument have precedence. Neither is it necessary for the old documentation to be revised and re-approved to align with the results from the new Stability Instrument.

6. CONCLUSIONS

This paper has described some of the problems relating to the complex issue of tanker damage stability which have eventually led to revisions to IMO's tanker codes [8] [12] [13] [16] [17] [18].

Although calculation of the damage stability of tankers is complex, the advent of ever-improving software and hardware has opened up the possibility of such complex calculations being undertaken quickly and economically on-board a ship (or at a shore station) within a reasonable period prior to the ship's departure using the actual loading condition, and if necessary, those to be encountered throughout a voyage. The performance of such calculations on-board is considered preferable to placing reliance upon shore-based stations, and hence the amended IMO Regulations promote the use of these improved computer systems (on-board Loading Instruments) in preference to using simplified manual and semi-manual methods for verifying compliance with damage stability regulations.

The preference of the UK's MCA is for the use of on-board stability Loading Instruments running Type 3 software on all tankers subject to the amended Regulations even though these still permit existing methods of verification to be retained on tankers constructed before 1st January 2016. New MCA guidelines in the form of MGN 611 (M) [2] have recently been produced in consultation with stakeholders, hence its publication should assist all stakeholders in applying the regulatory amendments and associated guidelines, and it will obviously be subject to regular review.

7. DISCLAIMER

The views expressed in this paper are those of the authors and do not necessarily represent those of the organisations with which they are affiliated and the professional institutions of which they are members.

8. REFERENCES

1. HUTCHINSON, K.W. and SCOTT, A.L. 'Tanker Damage Stability: What are the Issues and Solutions', *International Conference on the Design and Operation of Tankers*, The Royal Institution of Naval Architects, Athens, Greece, 8th and 9th June 2011. ISBN: 978-1-905040-85-8.

2. MGN 611 (M) Damage Stability: Alternative verification method for tankers - UK interpretations and procedures, Maritime and Coastguard Agency, Southampton, United Kingdom, 18th July 2019. https://www.gov.uk/government/publications/mgn-611-m-damage-stability-alternative-verification-method-for-tankers

3. HUTCHINSON, K.W. and SCOTT, A.L. 'Tanker damage stability: historical problems', *The Naval Architect*, The Royal Institution of Naval Architects, London, United Kingdom, March 2018, pages 34 to 39. ISSN: 0306-0209

4. HUTCHINSON, K.W. and SCOTT, A.L. 'Tanker damage stability: a solution to the historical problems', *The Naval Architect*, The Royal Institution of Naval Architects, London, United Kingdom, June 2018, pages 34 and 35. ISSN: 0306-0209

5. HUTCHINSON, K.W. and SCOTT, A.L. 'Damage stability considerations for chemical and product tankers', *The Naval Architect*, The Royal Institution of Naval Architects, London, United Kingdom, November 2018, pages 37 to 40. ISSN: 0306-0209.

6. HUTCHINSON, K.W. and SCOTT, A.L. 'Tanker damage stability: proposed solutions', *The Naval Architect*, The Royal Institution of Naval Architects, London, United Kingdom, September 2019, pages 60 to 64. ISSN: 0306-0209.

7. HUTCHINSON, K.W., SCOTT, A.L., WRIGHT, P.N.H., WOODWARD, M.D. and DOWNES, J., 'Consideration of Damage to Ships from Conceptual Design to Operation: The Implications of Recent and Potential Future Regulations regarding Application, Impact and Education'. *The Damaged Ship International Conference*, The Royal Institution of Naval Architects, London, United Kingdom, 26th to 27th January 2011. ISBN: 978-1-905040-79-7

8. MARPOL Consolidated Edition 2017 *Articles, Protocols, Annexes, Unified Interpretations of the International Convention for the Prevention of Pollution from Ships, 1973, as modified by the Protocol of 1978 relating thereto*, International Maritime Organization, London, United Kingdom, 2017. ISBN 978-92-801-1657-1

9. SOLAS Consolidated Edition 2014 *Consolidated text of the International Convention for Safety of Life at Sea, 1974, and its Protocol of 1988: articles, annexes and certificates. Incorporating all amendments in effect from 1 July 2014,* International Maritime Organization, London, United Kingdom, 2014. ISBN 978-92-801-1594-9

10. Resolution MSC.421(98) *Amendments to the International Convention for the Safety of Life at Sea, 1974, as amended*, Maritime Safety Committee, International Maritime Organisation, London, United Kingdom, 15th June, 2017.

11. Resolution MSC.436(99) *Amendments to the International Convention for the Safety of Life at Sea, 1974, as amended*, Maritime Safety Committee, International Maritime Organisation, London, United Kingdom, 24th May, 2018.

12. IBC Code 2016 Edition *International Code for the Construction and Equipment of Ships Carrying Dangerous Chemicals in Bulk*, International Maritime Organization, London, United Kingdom, 2016. ISBN 978-92-801-1595-6

13. BCH Code 2008 Edition *Code for the Construction and Equipment of Ships Carrying Dangerous Chemicals in Bulk*, International Maritime Organization, London, United Kingdom, 2009. ISBN 978-92-801-1509-3

14. HUTCHINSON, K.W. and SCOTT, A.L. 'Stability of RO-PAX Ships in European Waters: SOLAS 2020 and the Stockholm Agreement', *International Conference on Sustainable and Safe Passenger Ships*, The Hellenic Institute of Marine Technology in association with the Royal Institution of Naval Architects, Athens, Greece, 4th March 2020. ISBN: 978-1-911649-04-5

15. HUTCHINSON, K.W. and SCOTT, A.L. 'SOLAS 2020 Damage Stability and beyond', *International Conference on The Damaged Ship V*, The Royal Institution of Naval Architects, London, United Kingdom, 11th March 2020. ISBN: 978-1-911649-05-2

16. IGC Code 2016 Edition *International Code for the Construction and Equipment of Ships Carrying Liquefied Gasses in Bulk*, International Maritime Organization, London, United Kingdom, 2016. ISBN 978-92-801-1631-1

17. GC Code 1983 Edition *Code for the Construction and Equipment of Ships Carrying Liquefied Gasses in Bulk*, International Maritime Organization, London, United Kingdom, 1983. ISBN 978-92-801-1165-1

18. EGC Code 1976 Edition *Code for Existing Ships Carrying Liquefied Gasses in Bulk*, International Maritime Organization, London, United Kingdom, 1976. ISBN 978-92-801-1051-7

19. Load Lines Consolidated Edition 2005 Edition *International Convention on Load Lines, 1966 and Protocol of 1988, as amended in 2003*, International Maritime Organization, London, United Kingdom, 2005. ISBN 92-801-4194-5

20. Load Lines, International Convention on Load Lines, 1966 and Protocol of 1988, as amended, Consolidated Edition 2005, Supplement December 2013, International Maritime Organization, London, United Kingdom, 2013. (Sales number IB701E)

21. SLF 52/INF.4 *Guidelines for Verification of Damage Stability Requirements for Tankers and Bulk Carriers - Guidelines for scope of damage stability verification on new oil tankers, chemical tankers and gas carriers — Submitted by the International Association of Classification Societies (IACS)*, International Maritime Organisation, London, United Kingdom, 20th November, 2009.

22. Recommendation No.110 Rev.1 *Guidance for Scope of Damage Stability Verification on new oil tankers, chemical tankers and gas carriers*, International Association of Classification Societies Ltd., United Kingdom, November 2010.

23. MODU Code Consolidated Edition 2009 *Code for the Construction and Equipment of Mobile Offshore Drilling Units*, International Maritime Organization, London, United Kingdom, 2010. ISBN 978-92-801-1525-3

24. *Paris CIC Tanker Damage Stability Guidance Notes*, Paris Memorandum of Understanding on Port State Control, 28th May 2010.

25. UR L5 Rev.3 *Onboard Computers for Stability Calculations*, International Association of Classification Societies Ltd., United Kingdom, June 2017.

26. SOLAS Consolidated Edition, 1992, *Consolidated text of the International Convention for Safety of Life at Sea, 1974, and its Protocol of 1988: articles, annexes and certificates. Incorporating all amendments up to and including the 1990 amendments (1991 amendments as an appendix),* International Maritime Organization, London, United Kingdom, 1992. ISBN 978-92-801-1594-5

27. HUTCHINSON, K.W. and SCOTT, A.L., 'SOLAS 2020 Passenger and Cargo Ship Damage Stability Regulations', *Proceedings of the International Conference on The Damaged Ship IV*, The Royal Institution of Naval Architects, London, United Kingdom, 16th and 17th May, 2018. ISBN: 978-1-909024-75-5

28. HUTCHINSON, K.W. and SCOTT, A.L., 'Damage stability: upcoming SOLAS 2020 regulations', The Naval Architect, The Royal Institution of Naval Architects, London, United Kingdom, September 2017. ISSN: 0306-0209

29. MSC.1/Circ.1461 *Guidelines for Verification of Damage Stability Requirements for Tankers*, International Maritime Organization, London, United Kingdom, 8th July 2013.

30. Consultation on alternative verification methods for tankers, Maritime and Coastguard Agency, Southampton, United Kingdom, 16th July 2019. https://www.gov.uk/government/consultations/consultation-on-alternative-verification-methods-for-tankers

9. AUTHORS BIOGRAPHIES

Keith Hutchinson BEng(Hons) CEng CMarEng FRINA FIMarEST FSNAME is a Senior Consultant in Whole Ship Design and Naval Architecture with Safinah Group, where he undertakes commercial and offshore marine consultancy for a range of clients globally. He is a graduate in Naval Architecture and Shipbuilding from both Sunderland Polytechnic and Newcastle University. Having worked in both Shipbuilding and Shiprepair yards on the River Tyne, he joined Swan Hunter Shipbuilders in 1990 as a Naval Architect and then in 1995 Armstrong Technology, which was acquired by Babcock in 2000, becoming a Senior then the Principal Naval Architect and finally overall Engineering Manager for the Tyneside Ship Design office until its closure in 2018 - he was also Chief Engineer for Commercial Marine and Subject Matter Expert for the Naval Architecture domain across Babcock. Keith has published over fifty technical papers and articles and been invited to give numerous public presentations on ship and offshore vessel design and performance. He sits on a variety of professional technical committees developing standards and guidance for the maritime industry globally and is active in several professional learned institutions and their branches.

Andrew Scott MBE BSc CEng FRINA is Policy Lead for Stability, Load Line and Tonnage within the Ship Standards Division of the Directorate of Maritime Safety and Standards at the Maritime and Coastguard Agency (MCA). He is based in their Southampton headquarters, United Kingdom, but now works mostly from his home in the North East of England. He is responsible for policy matters relating to stability, load line and tonnage and was until recently co-ordinator of the International Maritime Organization's (IMO) Ship Design and Construction (SDC) Correspondence Group covering the latest updates to SOLAS 2009 and on issues relating to Roll-On / Roll-Off (RO-RO) stability. He has led several MCA-sponsored Research Programmes into the impact of the 2009 SOLAS amendments on ship design most specifically on the issue of Water-on-Deck (Stockholm Agreement). Andrew graduated from the University of Newcastle upon Tyne in 1969 in Naval Architecture and Shipbuilding and joined Lloyds Register of Shipping (LRS). In 1979 he moved to the British Ship Research Association (BSRA), later BMT, in Wallsend in 1979 as a Research Officer working in consultancy, particularly RO-PAX ships, before joining the MCA 2003.

PAST, PRESENT AND FUTURE OF DAMAGE STABILITY CALCULATIONS

R. Pérez, SENER Ingeniería y Sistemas, Spain

SUMMARY

Historically, damage stability calculations relied on manual calculations, often tied to a specific system of measurement. Some of these very old equations continue to be used in naval architecture books today. However, the advent of the towing tank allows much more complex analysis.
Damage stability calculations are much more complicated than other naval architecture calculations, as for example, intact stability. Software utilizing numerical methods are typically employed because the areas and volumes can quickly become tedious and long to compute using other methods.
Stability tests in damage conditions, as in shipyards, technical offices or universities, are performed with software packages that starting from conceptual design information are able to quickly compute the required data.
This paper focuses on evaluating old and new methods of damaged stability calculations. From the times of Pierre Bouguer and Fredrik Henrik auf Chapman until the latest developments in shipbuilding CAD tools.

NOMENCLATURE

CAD	Computer Aided Design.
GM	Permissible metacentric height (m).
GZ	Righting arm (m).
HARDER	Harmonization of Rules and Design Rationale.
IMO	International Maritime Organisation.
KG	Permissible height of the centre of gravity (m).
MSC	Maritime Safety Committee.
OSV	Offshore Service Vessel.
PSV	Platform Supply Vessels.
Ro-Ro	Roll On-Roll Off ship.
SOLAS	International Convention for the Safety of Life at Sea.
SPS	Special Purpose Ships.

1. INTRODUCTION

Naval architecture calculations have been used for hundreds of years, and damage stability was an important portion of them. Traditionally, damage stability calculations for ships relied on manual calculations, often tied to a specific system of measurement. Some of these very old equations continue to be used in naval architecture books today. Nevertheless, the appearance of the towing tank allowed much more challenging investigations. Damage stability calculations are much more problematical and complex than other calculations, as for example, intact stability (*Pérez, 2018*).
Shipbuilding *Computer Aided Design* (CAD) tools employ numerical methods, because the volumes and areas become tedious and time-consuming to compute using other methods. Stability tests in damage conditions, as in shipyards, technical offices or universities, are performed with software packages that starting from conceptual design information are able to quickly compute the required data. This paper focuses on evaluating old and new methods of damaged stability calculations in the shipbuilding *CAD* tools.

2. HISTORICAL BACKGROUND AS REGARDS SHIP STABILITY

Ship severity practical applications started around mid-18th century when the concept of metacentric height and righting arm were introduced for the first time by ***Pierre Bouguer (1698-1758)***, who published the first treatise of naval architecture in 1746, *Traité du navire* (see its title page in the *figure 1*), which among other achievements first explained the use of the metacentre as a measure of ships´ stability.

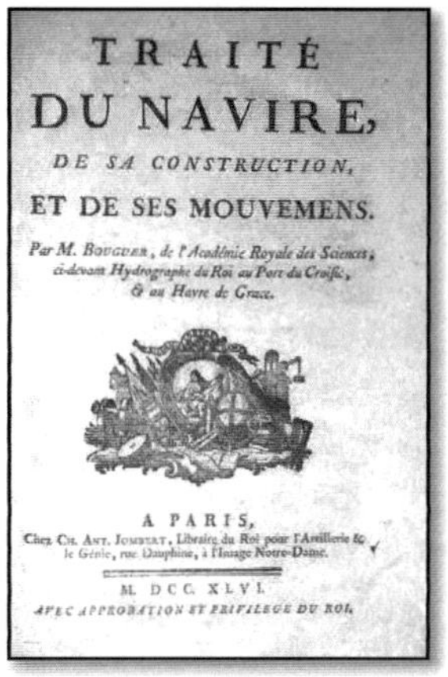

Figure 1. Cover page of Traité du navire. The first treatise on Naval Architecture published by Pierre Bouguer in 1746

It is worth it to mention that the first true stability criterion, based on metacentric height and freeboard height, was developed by ***Fredrik Henrik auf Chapman (1721-1808)*** during the same period.
Chapman is credited as the first person to apply scientific methods to shipbuilding and is the author of *Architectura Navalis Mercatoria* in 1768. In 1765 Chapman applied for permission from his work as a chief naval builder for the *Archipelago Fleet* at the naval base at Seaborg to work on

the book. However, it was the *Tractat om Skepps-byggeriet* (see its cover page in *figure 2*), translated in English as *Treatise on shipbuilding*, the crucial book who became the author as one of the leading experts on shipbuilding in the world.

TRACTAT
OM
SKEPPS-BYGGERIET,
ARCHITECTURA
NAVALIS MERCATORIA &c.
STOCKHOLM,

Figure 2. Title page of Tractat om Skepps-byggeriet. Published in 1775 and became Chapman one of the leading experts in the shipbuilding sector

In the 1816 century ***Mark Beaufoy (1764-1827)*** published a brilliant paper based on his experimental work entitle *On the Stability of Vessels*. The paper contains drawings of his new device (see an illustration from the book in *figure 3*), showing a model, he tested 23 different hulls, form being subjected to a controlled heeling force with a plumb bob and scale to measure the inclination.

Figure 3. Illustration of On the Stability of Vessels. Published in 1816 by Mark Beaufoy

Moving forward on the 19th century, ***William Henry White (1845-1913)*** worked on the design of the battleships *Royal Sovereign* (8 sister ships). He wrote in 1877 *A Manual of Naval Architecture: For the use of Officers of the Royal Navy, Shipbuilders* (see the cover page in *figure 4*). He proposed an increase in freeboard to prevent flooding, and carried out a comprehensive stability study for new hulls.

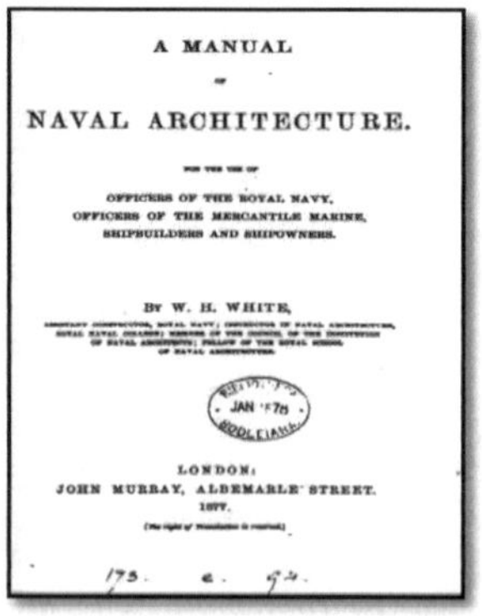
A MANUAL
OF
NAVAL ARCHITECTURE.
FOR THE USE OF
OFFICERS OF THE ROYAL NAVY,
OFFICERS OF THE MERCANTILE MARINE,
SHIPBUILDERS AND SHIPOWNERS.
BY W. H. WHITE,
LONDON:
JOHN MURRAY, ALBEMARLE STREET.
1877.

Figure 4. Front page of A Manual of Naval Architecture. Published in 1877 by William Henry White

Later on, ***Aleksei Nicolaevitch Krylov (1863-1945)*** wrote many papers and books on a wide range of topics, including shipbuilding. He published theories as regards stability. His floodability tables have been used worldwide.
The first damage stability regulations were defined in 1889 during the *International Maritime Conference* in Washington DC.
During the 20th century, several International Conferences have been celebrated as regards damage stability. In the following *table 1*, a brief description of each of the conferences has been provided.

Table 1. Historical timeline of the SOLAS Conferences

NAME	DESCRIPTION
SOLAS 1914	Signed by 13 countries on January 1914. Ratified only by five. Eventually interrupted by WWI and did not enter into force as intended on July 1915.
SOLAS 1929	New floodable length, compartment and subdivision requirements. Asymmetric subdivision to the satisfaction of the Administration.
SOLAS 1948	Minimum damage extent: longitudinal: 3.05 m ... 10.67 m; transverse: B/5; and vertical: top of double bottom to margin line. Deterministic criteria for maximum heel of 7° and no margin line immersion at final equilibrium.
SOLAS 1960	First major achievement by IMO. Signed on June 1960. Entered into force on May 1965
SOLAS 1974	A new type of convention, a convention which is amended instead of being replaced. Entered into force on May 1980. *SOLAS 1974*, as amended, is what we still use today.
SOLAS 1990	Resolution *MSC.12(56)*, adopted in October 1988. Entered into force April 1990.
SOLAS 1995	7 northwest European countries concluded a regional agreement, Stockholm Agreement
SOLAS 2009	Probabilistic damage stability requirements for both cargo and passenger ships.
SOLAS 2020	*MSC* adopted the updated regulation in Resolution *MSC.421(98)*. Restores the changed definitions on how to measure draught and trim in *SOLAS*.

It is well known that most of this conferences raised after a serious accident on the Sea. Some of the accidents are listed below *(Perez, 2018)*:

- **Titanic (1912).** Crashed with an iceberg during her maiden voyage from Southampton to New York and sank in the North Atlantic on April 1912. 1,500 persons lost their lives.
- **Empress of Ireland (1914).** Crashed into by *SS Storstadt* during thick fog in the Saint Lawrence River, Canada. Sank in the early hours of May 1914. 1012 persons lost their lives. No time to close watertight doors and the water entered through open portholes, some of them close to the waterline.
- **Andrea Doria (1956).** Crashed into by *SS Stockholm* outside Massachusetts. Sank on July 1956. 46 persons dead. Stayed afloat over 11 hours after the incident.
- **Herald of Free Enterprise (1987).** A *Ro-Ro* ferry that capsized just after leaving port. Sank on a sandbank in the evening on March 1987. 193 persons lost their lives.
- **Estonia (1994).** A ferry who capsized after the bow visor broke off in strong weather. Sank on September 1994 in the Baltic Sea. 852 persons on-board lost their lives.
- **Express Samina (2000).** Another ferry that crashed against the reefs in the Aegean Sea. Main damage above the waterline, but a stabilizer fin cut through the hull below the waterline. Watertight doors open and capsized and sank on September 2000. 81 persons onboard dead.
- **Costa Concordia (2012).** A cruise ship capsized after hitting a rock off Isola del Giglio, on the western coast of Italy. The evacuation took over six hours and of the 3,229 passengers and 1,023 crew known to have been aboard, 32 died.

A special mention, in this brief summary of the damage stability regulation history, it is for Jaakko Rahola. He published his doctoral thesis, at the *Technical University of Finland*, on May 1939. Entitled *The Judging of the Stability of Ships and the Determination of the Minimum Amount of Stability*, he especially considered the vessels navigating in Finnish waters, presenting the minimum stability using upper (the *GZ* curve) and lower curve (the dynamic stability curve).

3. DAMAGE STABILITY CALCULATIONS DURING THE 20th CENTURY

The first tools to evaluate damage stability were developed around 1970's, including flooding conditions and damage stability evaluation for defined loading conditions (see an example in *figure 5*).

Damage stability calculation is a subject for continuous increasing need. This the reason to have continuous improvements and upgrades on the *CAD* tools, as for example this is the main cause to have the *International Maritime Organisation* (IMO) probabilistic damage calculations, probabilistic damage assessment to handle new rules for cargo and passenger vessels, calculations for warships and other functionalities/ calculations which are carried out by most of the *CAD* Systems.

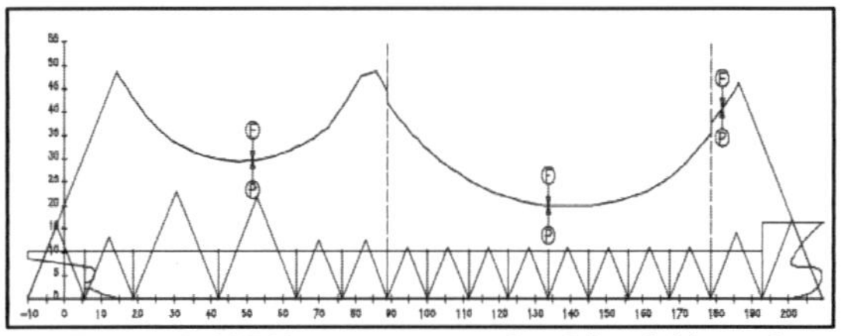

Figure 5. 1970's output example of floodable and permissible length curves (Perez & Muñoz, 2015)

It is also well known that there are two applicable damage stability calculation methods for vessels: deterministic and probabilistic damage stability. The correct/applicable method depends on the ship to be analysed. The probabilistic method, which is based on damage statistics in terms of the magnitude and position of the damage on previously damaged ships, is considered to be more reasonable than the deterministic approach.

The improvement of probabilistic damage stability regulations started in the late 1960s, and almost a decade later, *Resolution A.265 (VIII)* of *SOLAS 1974* entered into force. This approach was an alternative to the deterministic procedure for dry cargo and passenger ships, which still was a fragment of the *SOLAS*.

Today's prevailing probabilistic damage stability regulations can be found in *SOLAS Chapter II-1, Part B-1*, and stems from a project named *HArmonization of Rules and DEsign Rationale* (HARDER).

The *HARDER* project began to gather huge number of collision and grounding data in 2001, with the goal of harmonizing the damage stability regulations for all type of vessels. All existing methodologies to probabilistic damage stability calculations were reconsidered and new mathematical formulations were suggested to the existing regulations. The final proposals were adopted in 2005 by the *Maritime Safety Committee* (MSC), an underlying committee of *IMO*, and the new regulations come into force in 2009.

These are relevant for dry cargo ships with a length above 80 meters and passenger ships built after 1st of January 2009. With the new probabilistic damage stability regulations in place, *IMO* actually declared that deterministic damage stability has no future (*Vassalos, 2014*).

Due to the *Code of Safety* for *Special Purpose Ships* (SPS), which was adopted for first time in 2008 by *IMO* Resolution *MSC.266(84)*, *SPS* are also covered by the probabilistic damage stability regulations.

If a vessel has *SPS notation 1*, it is accepted as passenger ship in accordance with the damage stability regulations as regards *SOLAS Chapter II-1, Part B-1*.

Platform Supply Vessels (PSV) are, for example, not covered by the *SPS Code*, which means that *PSV* are excluded from the probabilistic damage stability regulations. *IMO Resolution MSC.235(82)* must then be applied, which sticks to a deterministic damage stability approach.

Offshore Service Vessels (OSV) are classified with *SPS* notation; more than ever, *OSV* are now involved in operations that needs more on-board distinct personnel. Wind farm operation service vessels could be an example of this type of tasks.

There are particular design advantages with the use of damage stability probabilistic approach. There are also other advantages using the traditional deterministic methodology. One of them is the straightforward calculation procedure that makes it possible to rapidly estimate the ship survivability.

The difficulty of the probabilistic damage stability regulations probably indicates how realistic and accurate the probabilistic approach is. Moreover, the probabilistic damage stability regulations most of times offer flexibility in the design of watertight arrangements which the deterministic damage stability regulations cannot compromise. With the deterministic damage stability regulations, the position of watertight bulkheads is defined by pre-defined damage extents. In the case with the probabilistic damage stability regulations, watertight bulkheads can be placed anywhere (*Djupvik et al, 2015*).

4. DETERMINISTIC DAMAGE STABILITY

Damage stability is obviously related to the permanency characteristics of a damaged ship, see *figure 6*. In *SOLAS*, a damage is defined as the three dimensional extent of the breach in the ship (*Perez & Riola, 2011a*). After injure, one or several compartments of the vessel may be filled with water, depending on the damage situation or position, the three dimensional damage magnitude and the watertight subdivision of the ship.

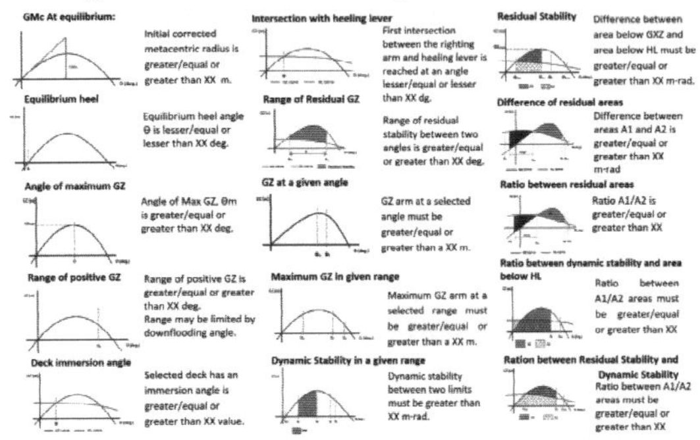

Figure 6. Stability criteria options

A space, compartment or tank, is defined by *SOLAS* as an on-board space within watertight bulkheads or boundaries, where volume is defined as a combination of *rooms*. A *room* is a part of the ship limited by bulkheads and decks, with a specific permeability (*IMO, 2008*). A compartment is said to be exposed to flooding when it is filled with water, and the water can move freely in and out of the compartment (*Patterson & Ridley, 2014*).

Traditionally, the purpose of a damage stability software is the calculation of the flooding conditions and the damage stability for the loading conditions defined previously. Normally, calculations can be performed according to the added weight method or to the lost buoyancy method.

- **Added Weight Method.** First step in this method is to calculate the volumes of the compartments flooded up to the initial situation, and then the corresponding modifications of the displacement and centre of gravity. This new obtained situation will not be an equilibrium situation, but the program will estimate the new drafts, trim and heel to obtain equilibrium. The equilibrium will be obtained when the buoyancy is equal to the displacement, and the centres of buoyancy and gravity are in the same perpendicular to the corresponding waterplane. This new estimated situation usually will not be the final equilibrium situation (but will be nearer than the previous one) and the flooded process will be performed again. The program will calculate the volumes of the compartments flooded up to the estimated situation, and then the corresponding modifications of the displacement and centre of gravity. After this the program checks the equilibrium (with the corresponding tolerances) and if the equilibrium has not been reached new estimation and a new iteration is performed. The process finishes when the equilibrium is reached. In each iteration, free surface correction of the compartments carrying liquid cargo and not flooded is done (when the initial situation is defined by a loading condition). For the calculation of the stability, a similar process is performed, but with a different equilibrium checking. In this case, for every heel angle, calculations are done with this fixed heel angle. The displacement will be the corresponding to the final situation after flooding. If the option of free communication with the sea has been selected, the displacement can be modified for every heel angle but the final values of the *GZ* curve, see *figure 7*, will be related to the equilibrium situation displacement.

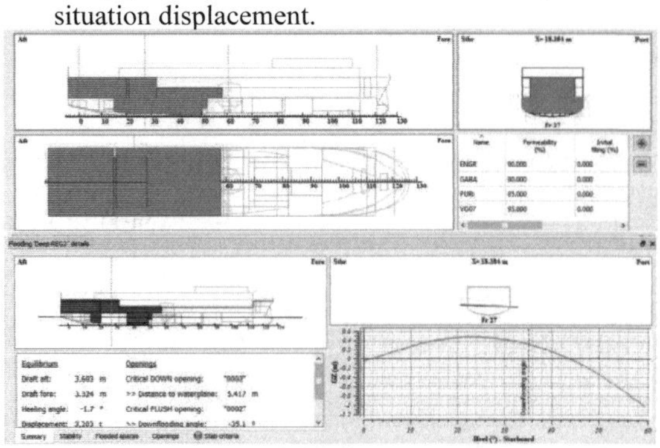

Figure 7. Damages and floodings appearance in a CAD System

- **Lost Buoyancy Method**. The displacement will be constant along all the calculations, except when there are flooded compartments carrying liquid cargo (this may happen when the initial situation is defined by a loading condition). In this case, the first step will be the loss of weights corresponding to this liquid cargoes. Then, the software will calculate the volumes of the compartments flooded up to the initial situation, and the corresponding modifications of the buoyancy, centre of buoyancy, waterplane characteristics etc. This new obtained situation will not be an equilibrium situation, but the program will estimate the new drafts trim and heel to obtain equilibrium. The equilibrium will be obtained when the buoyancy is equal to the displacement, and the centres of buoyancy and gravity are in the same perpendicular to the corresponding waterplane. This new estimated situation usually will not be the final equilibrium situation (but will be nearer than the previous one) and the flooded process will be performed again. The program will calculate the volumes of the compartments flooded up to the estimated situation, and the corresponding modifications of the buoyancy, centre of buoyancy, waterplane characteristics, etc. After this, the program checks the equilibrium (with the corresponding tolerances) and if the equilibrium has not been reached a new estimation and a new iteration is performed. The process finishes when the equilibrium is reached (the difference between the buoyancy and displacement are under defined tolerance and the same occurs with the residual arm). In each iteration, free surface correction of the compartments carrying liquid cargo and not flooded is done (when the initial situation is defined by a loading condition). For the calculation of the stability, a similar process is performed, but with a different equilibrium checking. In this case, for every heel angle, calculations are done with this fixed heel angle. The displacement is the intact displacement or modified by the lost liquids. To reach the equilibrium, buoyancy must be equal to this displacement and the centres of buoyancy and gravity are in the same perpendicular to the waterplane, but only in the longitudinal projection (free trimming). In the transversal projection the arm between these two forces will be the *GZ* arm for this heel.

The calculation of intermediate steps of flooding can be done in two different ways.

- **Further Lineal Method**. The engineer/designer must define a set of decreasing distances. These distances mean the parallel distance the compartment internal flotation is going to reach in relation to the previous stage external equilibrium flotation. In fact, the programme calculates the real equilibrium if it had entered a certain quantity of liquid up to this internal flotation and this internal flotation remains without entering more sea water at all. If the space is above the flotation, this space will be emptied up to the previous equilibrium flotation.
- **Sequential Method**. This method of calculation requires the designer to enter, for each compartment, how many steps are going to be studied up to final external flotation and in which step of the global number of steps begin the flooding. The programme will calculate this global number of steps. For instance, if compartment *A* is going to be flooded in three *steps* starting in *stage 2*, and compartment *B* in two *steps* beginning in *stage number 1*, it means that the calculation needs four *global steps*. When one space reaches its partial maximum number of steps, the programme, in any following step, will search the equilibrium in such a way that the internal flotation of the space was the external one. For the rest of stages the programme will calculate the volume proportionally to the number of steps of the space and the density of the liquid inside the flooded space will take into account the keys of replacement selected by the naval architect. When the initial internal level is above the external flotation the emptying out also is made sequentially.

Ship openings are those points where the watertight integrity of the ship is broken, and which are not fitted with permanent means of watertight closing. Downflooding will occur through these points if they become immersed in an equilibrium flotation.

Openings can have the following levels of watertightness:

- **Reference points**. They are points defined by the designer in order to obtain only information about their final position and the heeling angle of immersion. They are alike watertight openings and therefore they have not any influence in stability calculations.
- **Downflooding points**. Protected openings with weathertight closure. These points will effect progressive flooding when immersed at equilibrium but do not limit the range of stability curve.
- **Quick flooding points**. Permanent unprotected opening. These points will effect progressive flooding when immersed at equilibrium or will limit the range of stability.

The *CAD* Systems have normally the possibility of calculating either the maximum *KG* or the minimum *GM* complying with the criteria selected by two different tools:

- **Limiting KG/GM curve calculation**. In this case is compulsory to define a set of initial condition giving a range of middle drafts, the number of drafts for which the maximum *KG* is going to be calculated, the trim value and, optionally, a *KG* value to start iterations. The result is one maximum *KG* per draft (or one minimum *GM* per draft). For each of the drafts to be considered, the program will calculate the equilibrium flooding condition and the corresponding stability curve for the different values of *KG*, beginning with the initial value of *KB* and never higher to the initial value of *KM*.

  The program checks the results with the different sub-criteria in which the criteria selected have been split. If the flooding situation with the *KG* complies with all the sub-criteria and in one of them (at least) in a range lower than 2% that value of *KG* is considered the maximum *KG*; if one of the sub-criteria is not complied with that value of *KG* is rejected, if the program cannot find different values of *KG* closer to the real maximum *KG* or if the number of iterations for *KG* is reached, the maximum of all *KG* that complies with all sub-criteria is selected.

- **Limiting *KG* calculation complying with stability criteria for a flooding condition whose initial situation is an intact loading condition or by *drafts* + *KG*.** This option makes the *CAD* tool to evaluate the flooding condition for different values of initial *KG* up to find the greater *KG* value complying with all the criteria or advising that the programme is unable to find that value.

The calculations for water on deck are made in two stages:

- **First Stage**. Calculation of the height of water on deck, and the point in which residual freeboard occurs. The residual freeboard is the minimum distance between the damaged *Ro-Ro* deck and the final waterline at the location of the damage without taking into account the additional effect of sea water accumulated on the damaged *Ro-Ro* deck.
- **Second Stage**. Calculation of stability and criteria checking. There are two possibilities of calculation:
  - The volume of assumed accumulated sea water is kept the same at all trim and heeling angles and is calculated for the height of sea water from *step 1* in the ship's upright position.
  - The amount of sea water is calculated on the basis of a water surface having a fixed height above the lowest point of the deck edge of the damaged compartment calculated in *step 1* or when the deck edge in way of the damaged compartment is submerged, then the calculation is based on a fixed height above the still water surface at all heel and trim angles.

In both methods the accumulated water is added as a liquid load with one common surface inside all compartments which are assumed flooded on the car deck.

When the engineer/designer calculates one flooding condition, righting arms could be different according to the method used either lost buoyancy or added weight, because they are calculated with different values of displacement. Nevertheless the heeling moment must be equal in both methods, so, to make comparable the results of righting arms, it's enough to refer the heeling moment to the same displacement. In this way the same results will be obtained and both methods are comparable. This is the same thing to do with the calculation of *GM*, as it appears, for example, in (*Perez & Riola, 2011b*).

5. PROBABILISTIC DAMAGE STABILITY CRITERIA

Nowadays, the probabilistic damage stability calculations are applied to most of the cargo and passenger vessels according to different regulations.

The probabilistic approach for damage stability of ships was introduced by the German Professor ***Kurt Wendel*** in the late 1950s.

Latest regulation about damage stability from the probabilistic point of view is *SOLAS 2009 chapter II-1* based on *MSC194(80)*.

80th session of the *IMO*, *MSC* the working group finalized a substantial revision of *SOLAS chapter II-1 part A, B* and *B-1* aiming at harmonized damage stability requirements for all ship types except tankers performed by means of a common probabilistic method.

The draft adopted at the *MSC 80* will apply to all new passenger vessels, *Ro-Ro* and cargo ships built on or after 1st January 2009. The application date chosen will allow ensuring consistency and compatibility with other *SOLAS* chapters and finalizing the explanatory notes in order to arrive at user friendly formulation and homogenous interpretation of the revised damage stability requirements.

But there are also former regulations for cargo and passenger vessels. For dry cargo ships, and before the 1st of February 1992, there was not any regulation about damage stability, except the reference, on the 1966 *Load Line Convention*, to the vessels with reduced *type B* freeboards, which were required to survive with residual stability after sustaining damages up to a maximum defined size that was not necessarily applied to all locations in the ship.

Since February 1st of 1992 it was added to *SOLAS*, as *Part B-1* of the *chapter II-1*, a new Regulation for subdivision and damage stability of cargo ships over *100 (m)* in length that applies to the ships of this type constructed on or after February 1st of 1992. Later, the Regulation was extended to include ships with length between 80 and 100 meters.

Damage stability of passenger ships had been considered in *SOLAS* (*part B* of the *chapter II-1*) long time before the adoption of the probabilistic methods of calculation. The eight *Assembly* of *IMO*, by *Resolution A.265 (VIII)* adopted a set of probabilistic regulations of subdivision and damage stability of passenger ships as an equivalent to and a total alternative to the requirements of *part B* of *chapter II* of the *International Convention* for the *SOLAS 1960* for passenger ships.

Deterministic damage stability criteria is all about guaranteeing the safety of a ship. As it has been shown during the paper, it has been the methodology used for a long time, and the deterministic damage stability regulations are still in practise today for several reasons. Firstly because the probabilistic damage stability regulations do not cover all types of ships. Secondly, despite the deterministic damage stability regulations' well-known traditionalism, it has a long track-record, and the engineers inclines to trust on well-proven methods.

In addition to the *SOLAS 1974* standard, which regulates ordinary passenger ships, *table 2* provides an overview of other *IMO* instruments that contain deterministic damage stability provisions (*Hjort & Olufsen, 2014*).

Table 2. IMO instruments containing deterministic damage stability provisions, besides of SOLAS 1974 (Hjort & Olufsen, 2014).

REGULATORY FRAMEWORK	APPLICATION AREA

ICCL-66	*Cargo ships and tankers with reduced freeboard.*
MARPOL-73/78	*Tankers carrying cargo oil.*
IBC	*Code Ships carrying dangerous chemicals in bulk.*
IGC	*Code Ships carrying liquefied gases in bulk.*
HSC	*Code High Speed Crafts.*

6. DETERMINISTIC VS PROBABILISTIC CRITERIA

At this moment, the community of Naval Architects is debating between the probabilistic and the deterministic methods. It is therefore necessary to define what a deterministic or a probabilistic method is. To verify the validity of a model it is necessary to deduct from it a certain number of hypotheses and then to corroborate them with observations of predicted results.
Deterministic models correspond to mathematical models designed on the assumption that the result of an experiment is determined by the conditions under which it is performed; stochastic models (probabilistic) are those for which the data is obtained through a sampling of probability distributions. This sample allows that uncertainty (which can be reduced if more data is collected) and variability are propagated from the model and demonstrated in the results of the model.
Are probabilistic and deterministic approaches comparable?

- From a mathematical point of view: in general no.
- From a theoretical point of view: limited way.
- From a practical point of view (analysis of real ships…): yes.

Table 3. Probabilistic vs. deterministic approaches

APPROACH	*PROBABILISTIC*	*DETERMINISTIC*
Damage Size	*Variable*	*Fixed*
Number of damaged spaces	*No pre determinate*	*Fixed*
Damage Cases	*Several/Free*	*Few/Fixed*

In *tables 3* and *4* it is possible to visualize the probabilistic and deterministic approaches, plus the damage criteria applicability.

Table 4. Damage criteria applicability

SHIPS	*CARRIERS*	*PASSENGER*
Before 2009	*Probabilistic (Old SOLAS)*	*Deterministic (SOLAS 90) Probabilistic (A.265)*
After 2009	*Probabilistic (new rules)*	*Probabilistic (new rules)*
Size	*Lpp ≥ 80 (m)*	*All*

Are different probabilistic approaches comparable?

- From a mathematical point of view: yes.
- From a theoretical point of view: yes.
- From a practical point of view (analysis of real ships…): yes.

Not all the civil ships must follow the probabilistic criteria. There are several groups of ships that do not follow the new criteria and still using deterministic criteria:

- Tankers, chemical ships,
- Ships with a reduced freeboard.
- Special craft and offshore ships.
- High speed crafts.

In summary, it could be said that a deterministic model assumes that the actual result is determined by the conditions under which the experiment takes place, however when a stochastic model is used, the experimental conditions determine only probabilistic behaviour (the probability distribution) of the observable results.

7. FUTURE OF DAMAGE STABILITY CALCULATION TOOLS

The shipbuilding industry has always been very traditional and seems to be always in the tail of the implementation of the improvements that in other industries have matured previously. This fact has a reasonable explanation in the difficulty that the marine industry has to convert investments into profits. The marine industry is not friend of risks, especially when the simple fact of building ships is a huge risk in itself. On the other hand, the shipbuilding industry collects, to a greater or lesser extent, all the other industries and it may be thought that what is good for the others must also be good for it.
The evaluation of the stability criteria can be considered as design rules. These rules are established by international agreements and published by the *IMO*. Having the evaluation of the criteria integrated in the *CAD* shall have a positive impact in terms of cost savings. Otherwise it will be necessary to export the forms into separate format and to make some rework in the specialized software. Most of the specialized software covers the majority of them, but sometimes it is necessary to add specific calculations made with parametric formula. Not all the software are enabled to address it.
The use of user mathematical formulae is a powerful tool for customizing the rules, as it can be seen in the *figure 8*.

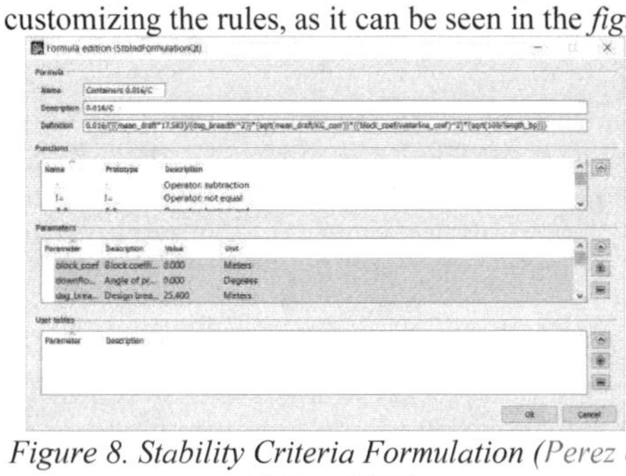

Figure 8. Stability Criteria Formulation (Perez & Muñoz, 2015)

Focused on the damage stability, *Artificial Intelligence* (AI) shall address important limitations among which are the lack of data or the confidentiality of the same. The marine industry has focused on immediate results, it is looking very quickly for a solution and does not store data and results in a systematic way that allows it to be used again in similar scenarios. The development of powerful algorithms requires that they can be applied in similar conditions with some recurrence. The data must be correctly structured and reasonably clean so that they can be used with advantage, *Muñoz et al. (2018)*. In the most successful cases there may be limited series of ships that have the same characteristics. It is difficult to find a systematic use of the data, but still there are interesting contributions that occur in the different phases of the damage stability life cycle. However, it is possible to find a recurrent pattern in the characteristics of the ship stability. So AI systems can assess if they are correctly defined.

8. CONCLUSIONS

CAD tools made damage stability calculations in accordance with the latest international conventions, regulations, rules and resolutions, which are continuously updated, even some *CAD* can define their own criteria.

CAD Systems enables accurate calculations of flooding conditions and damage stability according to the deterministic and probabilistic methods, taking into account the intact stability conditions, and associates the spaces to the geometric subdivision defined.

The short term future in the software development will be the calculation of the longitudinal strength after damage to know the situation after grounding, collision, etc. One of the features currently under development is the calculation of the effect of the crossfloodings pipes between compartments. The application will be able to calculate the effect of the flooding of some compartments including the connection among them. This will allow not only to have a more realistic knowledge of the behaviour of the vessel after the damage but also to identify actions to get equalization of the ship by flooding opposed compartments. These calculations will also include, for example, equalization time.

Future versions of these tools will incorporate the analysis of dynamic behaviour of the damaged ship in order to know the response of the ship to special sea conditions. The applications will also be able to predict the ship behaviour under corrective actions by the master, so supporting him about how to deal in a damage, what kind of actions would delay the capsizing, or even would avoid it. It is recognized that the current regulations are still not perfect, there is a lot of room for improvement, so in the future it is expected to have more advanced methods such as time domain simulation, and *CAD* tools which make the calculations easier.

With or without debate, the truth is that AI is present day by day in our environment, and its development will be growing. We will see AI in applications that we would not have imagined years ago. Musical compositions that are the most listened to can be created with AI or the most curious and tasty cooking recipes, intelligent cars without a driver or the best doctor capable of correcting diagnosis and treatment. But in spite of all these advances, we will continue to need intelligent people, people who are smarter than machines and who are ahead of them, because there are capacities of the human being who can never be embedded in an artificial intelligence.

The future innovations of these damage stability *CAD* tools offer a hopeful future for the shipbuilding Industry.

9. REFERENCES

1. DJUPVIK, OM.; AANONDSEN, SA. and ASBJØRNSLETT, BE., *'Probabilistic Damage Stability - Maximizing the Attained Index by Analyzing the Effects of Changes in the Arrangement for Offshore Vessels'*. NTNU, 2015.
2. HJORT, G. and OLUFSEN, O., *'Probabilistic damage stability'*. DNV GL AS, 2014.
3. IMO. Resolution MSC.281(85), *'Explanatory notes to the SOLAS Chapter II-1, Subdivision and damage stability regulations, Part B'*. International Maritime Organization, 2018.
4. LEWIS, E., *'Principles of Naval Architecture'*. Second Revision ed. Vol. I. Boston: The Society of Naval Architects and Marine Engineers, 1988.
5. MUÑOZ, JA.; PÉREZ, R. and GUTIERREZ, JR., *'Design Rules Evaluation through technologies of treatment of Big Data'*, 17th COMPIT, Pavone, 2018.
6. PATTERSON, CJ. and RIDLEY, JD., *'Ship stability, powering and resistance'*. Vol. 13. London: Adlard Coles Nautical, 2014.
7. PEREZ, R., *'The damage stability calculations evolution from the CAD point of view'*. DAMAGED SHIP IV. pp. 63-70. London, United Kingdom, 2018.
8. PEREZ, R. and MUÑOZ, JA., *'The Next Step for Damage Stability Calculations'*. DAMAGED SHIP III. pp 33-42. London, United Kingdom, 2015.
9. PEREZ, R. and RIOLA, JM., *'Case study of damage stability criteria of Merchant vessels and Warships'*. Damaged Ship International Conference. 26-27 January. London, UK, 2011a.
10. PEREZ, R. and RIOLA, JM., *'Damage Stability Criteria in Aircraft Carriers'*. Journal of Marine Technology and Environment, Vol. 1, 2011b.
11. VASSALOS, D., *'Damage stability and survivability-nailing passenger ship safety problems'*. Ships and Offshore Structures. 9(3), 237-256, 2014.

10. AUTHORS BIOGRAPHY

Rodrigo Perez Fernandez holds a PhD in Marine Engineering. He is head of the Naval Shipbuilding Area at SENER. He is also professor in several subjects at the Marine Engineering School of the Technical University of Madrid.

THE DIGITALIZATION OF NAVIGATION: EXAMINING THE ACCIDENT AND AFTERMATH OF US NAVY DESTROYER JOHN S. MCCAIN

S. C. Mallam, University of South-Eastern Norway, Norway
K. Nordby, The Oslo School of Architecture and Design, Norway
S. O. Johnsen, SINTEF Digital, Norway
F. B. Bjørneseth, Norwegian University of Science and Technology; Kongsberg Maritime CM AS, Norway

SUMMARY

A ship's bridge is a complex work environment where navigation crew must manage a plethora of interactions and tasks between people, technologies and systems both internal and external to the ship. In the 2017 collision between the US Navy Destroyer *John S McCain* and tanker *Alnic MC* in the Singapore Strait it was found that the design of the digital touch-screen steering and thrust control system increased the likelihood of operator errors, contributing to events leading to the accident. Furthermore, as a result, the US Navy announced that it planned to remove all touchscreen throttle and helm controls across their fleet in favour of physical controls. This paper discusses maritime digitalization through the lens of this case study in order to map out current issues, lessons learned and future development areas relating to digital applications for future bridge design and navigation operations.

NOMENCLATURE

GUI	Graphical User Interface
HMI	Human-Machine Interface
IBNS	Integrated Bridge and Navigation System
MBS	Multi-Vendor Bridge System
NTSB	National Transportation Safety Board
ODES	OpenBridge Design System
RPM	Revolutions Per Minute
SCC	Ship Control Console
SMACS	Sensemaking in Safety-Critical Situations
TSIB	Transport Safety Investigation Bureau
US	United States
USFFC	United States Fleet Forces Command

1. INTRODUCTION

Ship navigation involves complex interactions between the human crew and technologies that are set within a larger organizational and environmental context, often referred to as a socio-technical system (Carayon, 2006; Koester, 2007). Socio-technical systems consist of dynamic relationships between interrelated variables sharing a joint purpose and goals (Dekker, 2005; Wilson, 2014). Modern workplaces are becoming increasingly complex with tightly coupled interactions between subsystems and particularly between human agents and automated technologies (Perrow, 1993).

Ship bridges consist of a plethora of equipment covering a range of functions and purposes for the successful navigation and operation of the vessel. Ship navigation is cognitively demanding (Grech et al., 2008) and crew are required to manage and interact with these systems through differing types of digital and analogue inputs and outputs (Nordby et al., 2019).

When outfitting a bridge with equipment different companies and product types are typically installed, each potentially having their own Human-Machine Interface (HMI) design philosophies, styles and specifications (Nordby et al., 2019). Thus, Multi-Vendor Bridge Systems (MBS) force crew to work with, and manage, potentially unintegrated and disparate systems. Inconsistency of HMIs can decrease efficiency and promote unintended errors in operation (Nielsen, 2014). Although maritime regulatory authorities have taken action in developing bridge HMI design guidelines to promote well-designed equipment and systems, a gap continues to exist in current maritime design regulations and practices (Mallam & Lundh, 2013; Nordby et al., 2019).

The ship bridge has evolved considerably over recent history. Technology dictates the types of tasks and competencies required by human operators, and as technologies evolve, so to do the human requirements of working within a system. New technologies, such as more sophisticated automation of navigational systems (e.g. electronic navigation and autopilot) and increasing digitalization continue to create new demands for navigators.

However, new technologies, typically introduced in an effort to increase efficiency and safety, can introduce unintended complexities and even unforeseen consequences on a system and its operations (Bainbridge, 1983; Woods & Dekker, 2000). A popularly held and pervasive concept is that humans are the root-cause of accidents, due to their unreliability in otherwise safe systems (Dekker, 2014). Specifically, within the maritime domain, it is widely cited that human error is the cause of eighty percent or more of maritime accidents (Allianz Global Corporate & Speciality, 2012; Baker & McCafferty, 2005). However, this is both disingenuous and misleading, as this perspective only accounts for the pointy-end direct causes of accidents (i.e. the human operator and their actions at the end of a chain of events), rather than more systemic issues and contributory factors which ultimately lead to the so-called human error

(Dekker, 2014; Reason, 1991). Johnsen et al. (2019) found that technology-centric implementation of bridge systems without sufficient user-centred design considerations created systems that did not support crew work tasks, contributing to increased stress, poor sensemaking and conditions that can lead to accidents. Mallam et al. (2015) found that crew make their own modifications to ship work environments in order to increase work efficiency and ease, however these after-build alterations can also encourage unsafe work practices.

In 2014 a new ship bridge, Unified Bridge, was introduced to the marketplace, where a complete holistic re-design of the ship bridge was introduced, including consoles, levers and maritime software applications utilizing a user-centred design approach giving recommendations on the placement of equipment and emphasis on clear HMI design (Bjørneseth et al., 2014). In 2019, after 5 years in the market research found that crews responded positively to the changes, findings the bridge and equipment well arranged and user friendly (Danielsen et al., 2019).

Thus, a systems perspective, and specifically a socio-technical perspective of complex systems is critical to understanding the interactions between the various subsystems and their interactions. Human Factors aims to analyse these interactions between humans and technologies in order to optimize the design of systems from a socio-technical perspective (Nielsen, 2014; Stanton et al., 2017).

1.1 PURPOSE

The purpose of this paper is to analyze the collision and aftermath of the United States (US) Navy Destroyer *John S McCain* and tanker *Alnic MC* that occurred in the Singapore Straight in August 2017 with a focus on HMI design. Several reviews and accident reports were released in the aftermath of this accident, including from the United States Fleet Forces Command (USFFC) in October 2017 (USFFC, 2017), the Transport Safety Investigation Bureau (TSIB) of Singapore in March 2018 (TSIB, 2018), and the National Transport Safety Board (NTSB) report released in June 2019 (NTSB, 2019).

The accident investigation found, amongst other contributing factors, the design of the digital touch-screen steering and thrust control system increased the likelihood of operator errors, contributing to events leading to the collision. In August 2019, the US Navy announced plans to remove all touchscreen throttle and helm controls and replace with physical controls, such as mechanical joysticks, buttons and levers, across their fleet.

We investigate the general trend of digitalization and highly automated systems onboard modern ship bridges and explore the relationship between physical and digital systems. This accident provides a relevant case study and input into how the maritime domain moves forward in automating and digitizing the bridge, and the relationship between physical and digital solutions in ship operations.

2. BACKGROUND

NOTE: for a full account of the unclassified findings please refer to the official Marine Accident report issued by the NTSB (NTSB, 2019) and the Final Report issued by the Singapore Ministry of Transport (TSIB, 2018).

2.1 ACCIDENT SYNOPSIS

In the early morning hours of August 21, 2017 the destroyer, *John S McCain* (henceforth known as the "destroyer") was overtaking the tanker *Alnic MC* (henceforth known as the "tanker") as both ships (amongst other traffic in the area) were sailing south-westerly in the Middle Channel passage of the Singapore Strait Traffic Separation Scheme. The crew of the destroyer perceived a loss of steering, and while attempting to remedy the perceived situation, unintentionally turned to port and collided with the tanker. The weather was deemed not to be a concern or influencing factor, nor were any technical issues on either ship with regards to the propulsion systems, rudders or steering gear.

Ten crew onboard the destroyer died, forty-eight were injured and accident damage estimated over one-hundred million US dollars. Nobody onboard the tanker reported injuries and the ship sustained approximately one quarter of a million US dollars in damages.

2.2 SUMMARY OF ACCIDENT FINDINGS

The NTSB report found that the perceived loss of steering by the crew of the destroyer was due to the helmsman unintentionally transferring control of steering from the helmstation to the lee helmstation. The unintentional transfer misled the destroyer's bridge team, placing their focus on a perceived malfunction of the steering system (even though it was working properly). The bridge team did not input starboard rudder orders required for maintaining the ship's intended heading. The port and starboard rudders were not ganged (control of the port shaft and steering was at the lee helmstation, while control of the starboard shaft remained at the helmstation). The mismatched throttles resulted in an accelerated rate of turn to port. The crew did not recognize this mismatch, having lost situational awareness, and resulted in the collision with the tanker.

2.2 (a) Contributory Factors

The NTSB deemed that the probable cause of the collision between the two vessels was "*operational oversight*", which was attributed to specific contributory factors mentioned within the accident report:

- Insufficient training of crew
- Inadequate bridge operating procedures
- Loss of crew situational awareness
- Failure to follow steering emergency procedures
- Poor equipment design
- Crew fatigue

2.2 (b) Identified Design-Related Issues

The NTSB (2019) accident report states that: "*The design of the John S McCain's touch-screen steering and thrust control system increased the likelihood of the operator errors that led to the collision*" (Finding 23). This is in reference to the Ship Control Console (SCC) flat-panel touch screens of the helmstation and lee helmstation (which were identical in design). The touch screen Graphical User Interface (GUI) allowed crew to enter steering and thrust commands (see Figure 1).

It was concluded that the transfer of thrust control was unnecessarily complex and that the touch-screen throttle controls removed tactile feedback for helmsmen, making it more difficult to identify unganged or mismatched throttles.

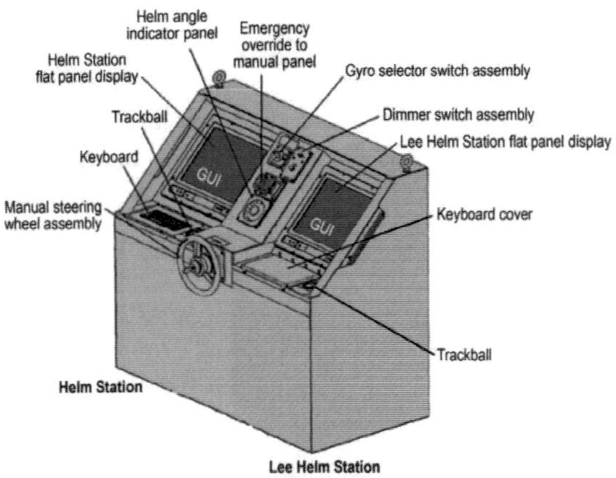

Figure 1. John S. McCain SCC (NTSB, 2019).

Both the NTSB (2019) and USFFC (2017) reports point to the touchscreens not being designed to industry best practices, or following a US Department of Defence approved standard for human engineering in marine systems (*ASTM International: Standard Practice for Human Engineering Design for Marine Systems, Equipment, and Facilities)* (ASTM International, 2013).

Furthermore, the USFFC (2017) report states that the modernization of their ships has lacked HMI reviews for safety critical equipment, leading to bad design. They stated that those responsible for installation had limited understanding of how to design for a ship's crew and their operations. This led to poor design choices in selection of control types, their layout and colour schemes (USFFC, 2017).

2.2 (c) Steering and Thrust Control Systems GUI

During operations at sea, where frequent changes to the thrust are not anticipated, control of both steering and thrust were normally located at the SCC helmstation. Here the helmsman steered the ship using the steering wheel and controlled thrust using the flat-panel display.

The steering control GUI of the Integrated Bridge and Navigation System (IBNS) was divided into three regions (see Figure 2): (i) the top region displays location of the station in control of steering and mode of steering in use; (ii) the middle region displays the gyrocompass repeater and rudder order/angle indicator; (iii) the bottom region displays which station is in control of thrust for each propeller, pitch and revolutions per minute (RPM). Each region and associated buttons opened further menus on the GUI display.

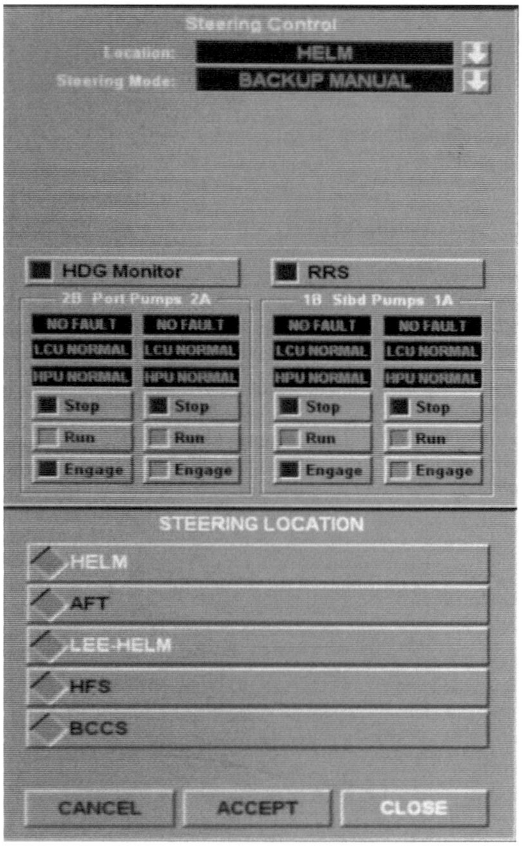

Figure 2. Steering control IBNS GUI (NTSB, 2019).

The thrust of each propeller shaft and speed could be controlled through the GUI displays including ganging and unganging the shafts (See Figure 3).

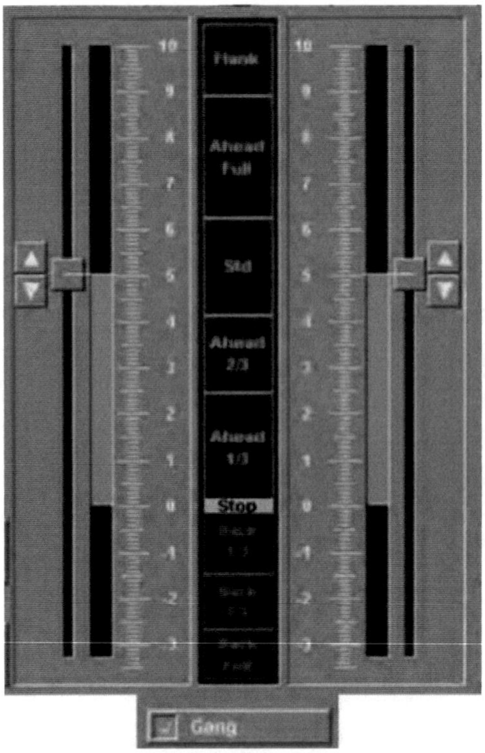

Figure 3. Port and starboard throttles and gang button (bottom) of the IBNS GUI (NTSB, 2019).

2.2 (d) Announcement of De-Digitalization Initiative

In August 2019, the US Navy announced that it would be reverting its destroyers from digital solutions, back to physical throttle and traditional helm control systems planned over the coming two years (2019-2021). This was based the accident findings, as well a fleet-wide surveys performed by the US Navy after the accident, with overwhelming preference found for physical controls over touchscreen systems (Eckstein, 2019).

3. DISCUSSION

This accident is a relevant case study for the modernization of bridge operating systems. It has gained much attention because of the issues related to the design of GUI systems as a contributing factor to the accident and the perceptions of an anti-digitalization initiative. However, this incident is an example of a typical issue related to Human Factors and socio-technical systems where new technologies, higher degrees of automation and increasing complexity of systems and operations impacts end-users and successful system functioning.

3.1 SYSTEM COMPLEXITY

This collision occurred in a heavily trafficked seaway (Singapore Strait) due to a complex set of variables and actions. Ultimately, the crew on the bridge of the *John S. McCain* lost situational awareness and misinterpreted the status of their propulsion displayed on the IBNS. Furthermore, there were GUI displays installed at various locations around the bridge and bridge wings, in visual sight by all of the primary watchstanders (at the time of the accident there were 14 people on the bridge, all of whom were qualified under the Navy's Personnel Qualification Standard). Due to this misinterpretation of the steering gear status (it had never actually failed) and subsequent tunnelling of focus to remedy a perceived, yet incorrect issue, none of the bridge team members noticed the mismatched shaft RPM increasing the ships rate of turn. Humans have a natural tendency to utilize skill- and rule-based behavioural patterns (Rasmussen, 1986). People try to match the current situation with previous experiences in an attempt to minimize cognitive effort used (Berman, 1997), thus resulting in a potential misperception and biasing of the current "real world" situation.

The destroyer had five modes to control steering, four using automatic steering (computer assisted manual mode was reported as the most commonly used mode), and a backup manual mode, without computer assistance. The IBNS was installed within a relatively short time-period (in 2016) of the accident (NTSB, 2019). According to both the USFFC report and the NTSB accident findings (Findings 24 and 25), crew may not have been fully trained or familiarized with the new IBNS system (NTSB, 2019; USFFC, 2017). The confusion over which steering mode the ship was in, which station had control, the status of steering gear, availability and clarity of information, communication between bridge team members, external communication and knowledge of the environment (i.e. surrounding ships and status), personal factors (i.e. fatigue, lack of technical or non-technical competencies) compounded the issue.

3.2 DIGITALIZATION ISN'T (NECESSARILY) THE PROBLEM

It is important to note that the SCC of the *John S McCain* (like most contemporary naval and merchant ships) is a combination of digital and physical inputs and outputs (see Figure 1). Furthermore, the US Navy's announcement to remove the digital steering and thruster controls are only two systems of a much larger work environment and set of equipment onboard.

The accident investigation found that the bridge crew (i) lacked knowledge of the newly installed system, and (ii) during operations lost an understanding of system status. The first issue points to a lack of training and knowledge of the system and its functions. Furthermore, the design failed to effectively present key information to crew, hindering their ability to extract the system's status quickly and accurately. This lack of system knowledge ultimately affected how the bridge team understood, analysed and reacted to this particular situation (i.e. the bridge team were focusing on the non-existent issue of a steering gear failure when there were no technical faults). This took their attention away from the critical information and key tasks required to avoid collision (i.e. poor sensemaking in a safety critical situation).

A lack of knowledge of how equipment functions is a threat to operators of highly automated, complex and safety-critical systems (Bainbridge, 1983). A balance must be struck between automation and human involvement (Kaber & Endsley, 1997). Human operators should have (at least) a basic level of knowledge of the expectations and underlying functions of automated and autonomous operations (Mallam et al., 2019). The accident between *John S McCain* and *Alnic MC* draws several parallels with the recent Boeing 737 Max 8 accidents. The operators in control of each respective complex, safety-critical transport medium (i.e. aircraft pilots and naval navigators) lacked sufficient training and knowledge of the underlying processes within the automated systems (Endsley, 2019; NTSB, 2019). As the investigators of the *John S McCain* accident found, the crew were not necessarily familiar with all functions of the relatively new bridge systems, and specifically, there were no transfer of control between bridge stations instructions in the Personnel Qualification Standard.

3.2 (a) Bad Design is Bad Design

Differences may exist between the *designers* of products, services or systems and the *users* of those products, services or systems (Hsu et al., 2000). Differences in backgrounds, experience, knowledge, expectations and demands between designers and users can create a disparity between what is created and how it is used (Mallam, 2016). Thus, if user needs are not well understood designers may not know how to develop effective solutions for users (Wilkinson & De Angeli, 2014). In engineering development, the traditional focus is primarily placed on technical functions and reliability, while user demands are lower in prioritization or not accounted for at all (Sun et al., 2013).

The NTSB (2019) report and USFFC (2017) review revealed that the touchscreens were not designed to best practices, did not follow approved human engineering guidelines, lacked HMI reviews, and according to the Program Executive Officer for Ships, were "*overly complex*" (Eckstein, 2019). Basic Human Factors and user-centred design principles strive for the system to support users and their demands (International Organization for Standardization, 2011). Good ergonomics is participatory and involves end-users throughout the design process (Lee et al., 2017; Wilson, 2014). The NTSB recommended that the US Navy incorporate human engineering design principles in their modernization of bridges (specifically mentioning ASTM, 2013). The US Navy has also begun to share lessons learned from accidents between designers, crew and education and training personnel (i.e. implementing participatory processes) (NTSB, 2019). Furthermore, the US Navy has stated that is has begun to consolidate responsibility and authority for bridge system modernizations (NTSB, 2019), referring to a more integrated approach for bridge design. All are examples of processes intended to improve the design of bridge systems.

The digitalized GUI that was found to be a contributory cause of this accident poses interesting questions for not only maritime applications but also modernization trends in general. Fleet-wide surveys post-accident found "*overwhelmingly*" that crew preferred analogue devices (Eckstein, 2019). This result is supported by the findings of Røed (2007), who found that interfaces using physical representations and perceptually rich interfaces may improve working conditions for the navigation team. The US Navy survey findings are a form of participatory input, eliciting expert end-user perspectives and knowledge (i.e. from US Navy personnel) in relation to bridge equipment design. However, it is important to note that these survey results may be biased and misleading, as they are based on poorly designed, poorly implemented digital solutions. In other words, digitalization alone might not be the primary reason for the desire to revert back to analogue devices, but rather poorly designed and integrated digital solutions. Just as poorly designed and integrated analogue solutions can create usability problems and compromise safe and efficient operations. Furthermore, there are differing demands on touch-based inputs and GUIs in moving environments (i.e. at sea). Motion can impede fine-motor control (Wertheim, 1998; Wertheim & Kistemaker, 1997). Thus, the motion of a vessel can impede the user's ability to repeatedly select small GUI targets (e.g. buttons, menus, sub-menu options) (Bjørneseth et al., 2012).

Regardless, digitalization may not be the best design solution for all bridge devices. More poignantly, we must better analyse the relationship between digital and physical devices on the bridge, and in complex control systems in general. A major criticism of the digital thrust and steering system was that in comparison to physical devices (which were reported to be much easier to see system status – i.e. direction and angle of a lever) digital interfaces hid information and was more difficult to cognitively decipher. An increasingly digitalized bridge has many threats, even apart from front-end design of HMIs. This can include the integration of multiple systems, their reliability and redundancy and cybersecurity. It is important in moving forward with bridge design that new and existing solutions are not developed in isolation but rather within the larger scope of system needs and goals.

Good maritime design standards and guidelines can help bridge these design gaps, however, they cannot be relied upon alone. The typical focus of maritime HMI design standards are on goal-based technical solutions (predominantly for physical design of bridges), which can be left open to interpretation, creating poor design and discrepancies across systems. Detailed standards and best practices that focus on interaction design and effectively support crews' cognitive demands are missing. A combination of not only effective design guidance but competent design teams with interaction and system design skills, end-user knowledge and Human Factors input can provide a powerful process for developing optimized bridge systems.

3.3 MOVING FORWARD

This paper discusses the collision between the US Navy Destroyer *John S McCain* and tanker *Alnic MC* through the lens of bridge equipment design. We highlight the major challenges of current bridge design processes and propose solutions to improve MBS, bridge digitalization and automation developed through two ongoing research projects: OpenBridge Design System (ODES) and Sensemaking in Safety-Critical Situations (SMACS).

3.3 (a) Improving Bridge Design and Consistency

Several issues have been identified in contributing to current bridge design deficiencies and inconsistency across bridge equipment. These issues are not isolated to this particular case, but typical across the maritime domain in naval and merchant ships. Firstly, current design guidelines for maritime equipment GUIs are both insufficient and rapidly aging (Nordby et al., 2019). This a concern because an increasing number of systems and functions are integrated on screens. In industries where digitization is more mature there is a greater convergence in design practices and patterns used for applications, which is currently only found to a limited degree in the maritime domain. There is a need to develop modern design guidelines for maritime systems (and specifically for digital systems) in order to ensure improved user interfaces across bridge equipment.

Secondly, ship's bridges are multi-vendor environments, creating design inconsistencies once equipment is installed into a single working environment. Currently, there are no methods for developing consistent design across all systems on a ship's bridge. As inconsistent designs can promote errors (Nielsen, 2014) there is a need to develop methods to address this issue.

Thirdly, the management between discrete physical interaction and digital GUIs must be better established. In the context of maritime operations, discrete physical interaction and graphical user interfaces both have their strength and weaknesses. Nordby & Morrison (2016) claim that due to the unique challenges and conditions of working at sea there is a need to consider extended support for multi-modal interaction, including physical controls, in ship bridges to enhance situational awareness. Nordby & Lurås (2015) explore how integrated bridge systems utilize both discrete physical controls and generic digital interfaces in order to derive benefits of both design approaches.

OpenBridge is an emerging design system developed to solve the three aforementioned challenges and improve design of digital maritime workplaces. OpenBridge design guidelines are open source to promote open innovation amongst maritime stakeholders. The initiative is built upon a user interface architecture seeking to streamline design of next generation multivendor maritime workplaces.

The user interface architecture organizes the bridge into the physical workplace, (digital) applications and an integration system, which is supported by user interface design guidelines aiming to improve HMI design and design consistency across systems (see Figure 4). Furthermore, Human Factors expertise and participatory processes, that elicit end-user knowledge and experience, are implemented throughout development cycles in order to optimize user-centred design solutions (Mallam 2016; NTSB, 2019). Both the design framework (i.e. OpenBridge design guideline and design standards for workplace hardware and applications) and the integration system (i.e. the methodology of implementation of design standards and user-centred approaches) are critical for success.

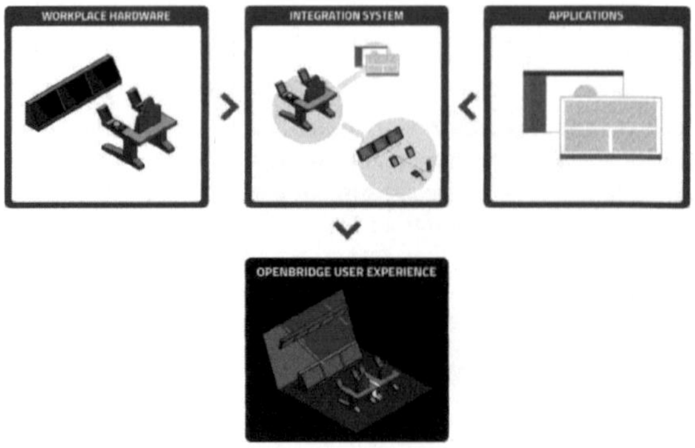

Figure 4. The three main levels of the user interface architecture (Nordby et al., 2019).

A core concept of OpenBridge is the integration of physical and digital solutions within the same system, leveraging the strengths of each interaction medium. Thus, the design architecture and integration process is key for successfully enabling multimodal user interfaces for both digital and physical solutions as part of a single work system.

3.3 (b) Improving Sensemaking on the Bridge

Sensemaking is defined as the dynamic, iterative process of observing, orienting, and acting in a social setting, thereby creating a shared understanding (Kilskar et al., 2019). Sensemaking on the bridge is a key prerequisite for performing the correct actions in safety critical operations (Johnsen et al., 2019). Sensemaking on the bridge is a complex social construct, dependent on training, procedures, responsibilities, environment, operational variables and the technology defining and supporting operations.

The SMACS project has found various areas that must be addressed in order to improve sensemaking for bridge crews. This includes the need for user-centred design principles (Danielsen & Lamvik, 2019; Johnsen et al., 2019); improved alarm systems and alarm guidelines (Danielsen et al., 2019; Hurlen et al., 2019); increased focus on training and training verification (Hurlen et al., 2019); a contemporary systems perspective in maritime accident analysis (Johnsen et al., 2019); adoption of best practices from other industries and high reliability organizations (Hurlen et al., 2019; Johnsen et al., 2019).

Ultimately, the design and integration of bridge equipment must support crew demands and work tasks. As we have discussed and demonstrated through this case study, design plays a crucial role in influencing people's perceptions of how they understand the world around them and thus, their actions taken. Through well-designed, well-integrated systems, we can reduce the negative effects of poor design and support the operators working in complex safety-critical environments.

4. CONCLUSIONS

The accident between *John S McCain* and *Alnic MC* has received much attention from the maritime domain, general media and technologists. It is the newest example of an old story: amongst other contributing factors, the combination of complex systems and functions, new technologies, automation, poorly designed and integrated systems, training and education and non-technical issues in operations led to an unintended, unanticipated accident.

Furthermore, this accident is just one example of a larger trend in recent years of collisions between naval and merchant ships (e.g. US Navy destroyer *USS Fitzgerald* collision with container ship *MV ACX Crystal* [2017]; Norwegian frigate *HNoMS Helge Ingstad* collision with tanker *MT Sola TS* [2018]). It is not within the scope of this paper to compare and contrast these accidents. However, it does raise questions over the culture and organizational factors of navies and naval operations in non-combative waters or situations that may have contributed to these accidents.

In the case of *John S McCain*, there were several identified factors that led to the accident, including insufficient training, inadequate operating procedures, poor equipment design and crew fatigue. The bridge crew failed to elicit key information, and even misinterpreted the situation due to the design of the SCC and its GUI for steering and thrust control. However, a root cause is not necessarily digitalization itself, but poorly designed and poorly implemented digital solutions within the complexity of highly automated systems. Digitalization has several advantages, particularly over the relatively long lifecycle of a ship. As ships age, inevitable equipment updates and retrofits will occur. Implementing digitalized solutions allows for far greater flexibility and adaption from a systems perspective. Research points towards tools and methodologies that support MBS design, increased HMI consistency and integration of bridge systems, including adopting the best aspects of both physical and digital solutions in the modernization of bridges and bridge operations. The debate between physical and digital is not an either/or decision, rather it is an issue of design quality and fit.

5. ACKNOWLEDGEMENTS

This work is completed for the OpenBridge Design System project, funded by the Norwegian Research Council (project no. 296151), OpenBridge consortium partners, and the Norwegian Research Council funded SMACS project - Sensemaking in safety-critical situations (project no. 267509).

6. REFERENCES

ALLIANZ GLOBAL CORPORATE & SPECIALTY, 'Safety and Shipping 1912-2012: From Titanic to Costa Concordia', *Allianz Global Corporate & Specialty*, 2012.

ASTM INTERNATIONAL, 'Standard Practice for Human Engineering Design for Marine Systems, Equipment, and Facilities (F1166-07)', *ASTM International*, 2013.

BAINBRIDGE, L., 'Ironies of automation', *Automatica*, 19(6): 775-779, 1983.

BAKER, C. C. & MCCAFFERTY, D. B., 'Accident Database Review of Human Element Concerns: What Do The Results Mean For Classification?', *In Proceedings of The Royal Institution of Naval Architects Conference on Human Factors in Ship Design, Safety and Operation*, 2005.

BERMAN, J., 'Design and Support for Abnormal Situations', *In Human Factors in Safety-critical Systems (Eds. F. Redmill & J. Rajan)*, 203-229, 1997.

BJØRNESETH, F. B., CLARKE, L., DUNLOP, M. & KOMANDUR, S., ''Towards an understanding of operator focus using eye-tracking in a safety-critical maritime setting', *In Proceedings of International Conference on Human Factors in Ship Design & Operation*, 2014.

BJØRNESETH, F. B., DUNLOP, M. D. & HORNECKER, E., 'Assessing the effectiveness of direct gesture interaction for a safety-critical maritime application', *International Journal of Human Computer Studies,* (70), 729-745, 2012.

CARAYON, P., 'Human factors of complex sociotechnical systems', *Applied Ergonomics*, 37(4), 525-535, 2006.

DANIELSEN, B. E., BJØRNESETH, F. B. & VIK, B., 'Chasing the end-user perspective in bridge design', *In Proceedings of Ergoship 2019*, 2019.

DANIELSEN, B. E. & LAMVIK, G., 'Making sense of bridge design: how seamanship may challenge technology-as-designed', *In Proceedings of ESREL 2019 - 29th International European Safety and Reliability Conference*, 2019.

DEKKER, S., 'The Field Guide to Understanding "Human Error"' (3rd ed.), *CRC Press*, 2014.

DEKKER, S. W., 'Why we need new accident models', *Lund University School of Aviation Technical Report*, 2005.

ECKSTEIN, M., 'Navy Reverting DDGs Back to Physical Throttles, After Fleet Rejects Touchscreen', *USNI News*, 2019.

ENDSLEY, M. R., 'Human Factors & Aviation Safety: Testimony to the United States House of Representatives Hearing on Boeing 737-Max8 Crashes – December 11, 2019', *Human Factors and Ergonomics Society*, 2019.

GRECH, M. R., HORBERRY, T. J. & KOESTER, T., 'Human Factors in the Maritime Domain', *CRC Press*, 2008.

HSU, S. H., CHUANG, M. C. & CHANG, C. C., 'A semantic differential study of designers' and users' product form perception', *International Journal of Industrial Ergonomics*, 25(4), 375-391, 2000.

HURLEN, L., SKJERVE, A. B. & BYE, A., 'Sensemaking in safety-critical situations. The challenges faced by dynamic positioning operators', *In Proceedings of ESREL 2019 - 29th International European Safety and Reliability Conference*, 2019.

INTERNATIONAL ORGANIZATION FOR STANDARDIZATION, 'Ergonomics – General approach, principles and concepts (ISO 26800:2011)', *International Organization for Standardization*, 2011.

JOHNSEN, S. O., KILSKAR, S. S. & DANIELSEN, B. E., 'Improvements in rules and regulations to support sensemaking in safety-critical maritime operations', *In Proceedings of ESREL 2019 - 29th International European Safety and Reliability Conference*, 2019.

KABER, D. B. & ENDSLEY, M. R., 'The combined effect of level of automation and adaptive automation on human performance with complex, dynamic control systems', *In Proceedings of the Human Factors and Ergonomics Society Annual Meeting*, 1997.

KILSKAR, S. S., DANIELSEN, B. E. & JOHNSEN, S. O., 'Sensemaking in Critical Situations and in Relation to Resilience—A Review', *ASCE-ASME Journal of Risk and Uncertainty in Engineering Systems, Part B: Mechanical Engineering*, 2019.

KOESTER, T., 'Terminology Work in Maritime Human Factors: Situations and Socio-Technical Systems', *Frydenlund*, 2007.

LEE, J. D., WICKENS, C. D., LIU, Y. & BOYLE, L. N., 'Designing for people: An introduction to human factors engineering', *CreateSpace*, 2017.

MALLAM, S. C., 'Distributed Participatory Design in Multidisciplinary Engineering Projects: Investigating a Sustainable Approach for Ship Design & Construction', *Chalmers University of Technology*, 2016.

MALLAM, S. C. & LUNDH, M., 'Ship Engine Control Room Design: Analysis of Current Human Factors & Ergonomics Regulations & Future Directions', *In Proceedings of the Human Factors and Ergonomics Society Annual Meeting,* 2013.

MALLAM, S. C., LUNDH, M. & MACKINNON, S. N., 'Integrating Human Factors & Ergonomics in Large-Scale Engineering Projects: Investigating a Practical Approach for Ship Design', *International Journal of Industrial Ergonomics*, 50, 62-72, 2015.

MALLAM, S. C., NAZIR, S. & SHARMA, A., 'The human element in future maritime operations – perceived impact of autonomous shipping', *Ergonomics,* 2019.

NATIONAL TRANSPORTATION SAFETY BOARD, 'Collision between US Navy Destroyer John S McCain and Tanker Alnic MC. Singapore Strait 5 Miles Northeast of Horsburgh Lighthouse August 21, 2017 (Marine Accident Report: NTSB/MAR-19/01 PB2019-100970)', *National Transportation Safety Board*, 2019.

NIELSEN, J., 'Coordinating user interfaces for consistency', *Elsevier*, 2014.

NORDBY, K. & LURÅS, S., 'Multimodal interaction for marine workplaces used as strategy to limit effect of situational impairment in demanding maritime operations', *In Proceedings of the International Conference on Marine Design*, 2015.

NORDBY, K., MALLAM, S. C. & LÜTZHÖFT, M., 'Open user interface architecture for digital multivendor ship bridge systems', WMU Journal of Maritime Affairs, 18(2), 297-318, 2019.

NORDBY, K. & MORRISON, A. D., 'Designing calm technology and peripheral interaction for offshore service vessels', *Personal and Ubiquitous Computing*, 20(4), 601-613, 2016.

PERROW, C., 'The Organizational Context of Human Factors Engineering', *Administrative Science Quarterly*, 28(4), 521-541, 1983.

RASMUSSEN, J., 'Information Processing and Human-Machine Interaction: An Approach to Cognitive Engineering', *Elsevier Science Inc.*, 1986.

REASON, J., 'Human Error', *Cambridge University Press,* 1991.

RØED, B. K., 'Designing for high-speed ships', *Norwegian University of Science and Technology*, 2007.

STANTON, N. A., SALMON, P. M., RAFFERTY, L. A., WALKER, G. H. & JENKINS, D. P., 'Human factors methods: a practical guide for engineering and design', *CRC Press*, 2017.

SUN, H., HOUSSIN, R., GARDONI, M. & DE BAUVROND, F., 'Integration of user behavior and product behaviour during the design phase: Software for behavioural design approach', *International Journal of Industrial Ergonomics*, 43(1), 100-114, 2013.

TRANSPORT SAFETY INVESTIGATION BUREAU, 'Safety Investigation Into Collision Between Alnic MC and The USS John S McCain In Singapore Territorial Waters on 21 August 2017 (MIB/MAI/CAS.021)', *Singapore Ministry of Transport*, 2018.

UNITED STATES FLEET FORCES COMMAND, 'Comprehensive Review of Recent Surface Force *Incidents', United States Fleet Forces Command*, 2017.

WERTHEIM, A. H., 'Working in a moving environment'. *Ergonomics,* 41(12), 1845–1858, 1998.

WERTHEIM, A. H. & KISTEMAKER, J. A., 'Task Performance During Simulated Ship Movements' (TNO-TM report TM-97-A014), *TNO Human Factors Research Institute*, 1997.

WILKINSON, C. R. & DE ANGELI, A., 'Applying user centred and participatory design approaches to commercial product development', *Design Studies,* 35(6), 614-631, 2014.

WILSON, J. R., 'Fundamentals of systems ergonomics/human factors', *Applied Ergonomics*, 45(1), 5-13, 2014.

WOODS, D. & DEKKER, S., 'Anticipating the effects of technological change: a new era of dynamics for human factors', *Theoretical Issues in Ergonomics Science*, 1(3), 272-282, 2000.

7. AUTHORS BIOGRAPHY

Steven C. Mallam is an Associate Professor in Maritime Human Factors in the Department of Maritime Operations at University of South-Eastern Norway. Steven's research focuses on improving human performance of individuals and teams in complex work systems through education, training and design solutions.

Kjetil Nordby is a Professor in Interaction Design with the Institute for Design at The Oslo School of Architecture and Design. Kjetil is leader of the Ocean Industries Concept Lab. Kjetil has extensive experience working as project leader, researcher, interaction and industrial designer for industry and academia.

Stig Ole Johnsen is a senior scientist at SINTEF in Norway, in the group safety and reliability. In addition he is managing the Human Factors in Control Network in Norway. His research focuses on human factors and safety and security of digitalization, automated and complex systems.

Frøy Birte Bjørneseth is an Associate Professor in Maritime Human Factors in the Department of Ocean Operations and Civil Engineering at the Norwegian University of Science and Technology in Ålesund, Norway. She is also a Principal Engineer in Human Factors and Control Centres in Kongsberg Maritime CM AS, a worldwide supplier of maritime equipment. Frøy Birte's research focuses on human factors in demanding maritime operations and holistic user centred design thinking in ship bridge and control centre environments.

IF YOU DOUBT THE VALUE OF SAFETY, TRY AN ACCIDENT

D. Vassalos and **D. Paterson,** University of Strathclyde, UK

SUMMARY

With the residual risk from damage stability limitations being so high, it makes no economic sense to invest on prevention, an obvious target for any safety-alert professional. What is the alternative then? Should we expect modern society, because of its vital dependence on shipping, to eventually accept the inevitability of mass death because of ship vulnerability to damage or instead consider a different, more pragmatic approach to maritime safety, in principle damage stability? Realising that the answer to this question is deterministic and one-sided, the EU maritime industry have launched, through funding from the EC, a large-scale Project (Flooding Accident Response – FLARE) aiming to establish a life-cycle risk-based regulatory framework for flooding risk management, applicable to new and existing vessels whilst considering passive and active means of risk reduction. This paper elaborates on these developments and highlights a recent innovation developed at the University of Strathclyde over the past 5 years that facilitates and supports such developments.

1. INTRODUCTION

With the residual risk from damage stability limitations being so high, it makes no economic sense to invest on prevention, an obvious target for any safety-alert professional. What is the alternative then? Should we expect modern society, because of its vital dependence on shipping, to eventually accept the inevitability of mass death because of ship vulnerability to damage or instead consider a different, more pragmatic approach to maritime safety, in principle damage stability? Realising that the answer to this question is deterministic and one-sided, the EU maritime industry have launched, through funding from the EC, a large-scale Project (Flooding Accident Response – FLARE) aiming to establish a life-cycle risk-based regulatory framework for flooding risk management, applicable to new and existing vessels whilst considering passive and active means of risk reduction. This paper focuses on innovation that has been developed at the University of Strathclyde over the past 5 years that facilitates and supports such developments.

2. BACKGROUND AND JUSTIIFICATION

Every time there is an accident with passenger ships, exposing their vulnerability to flooding as a result of collision/grounding accidents, societal outcry follows and industry and academia "buckle up", delving for design improvements to address the Achilles heel of this ship type, namely inadequate damage stability. However, any such improvements are targeting mainly newbuildings, which comprise a small minority of the existing fleet. Therefore, state-of-the-art knowledge on damage stability is all but wasted, scratching only the surface of the problem and leaving thousands of ships with severe vulnerability, that is likely to lead to further (unacceptably high) loss of life. This problem is exacerbated still further, today more rapidly, as the pace of scientific and technological developments is unrelenting, raising understanding and capability to address damage stability improvements of newbuildings cost-effectively, in ways not previously considered. As a result, SOLAS is becoming progressively less relevant and unable to keep up with this pace of development. This has led to gaps and pitfalls, which not only undermine safety but inhibit progress.

However, lack of retrospectively applied legislation (supported by what is commonly known as the Grandfather Clause) is not the only reason for damage stability problems with passenger ships. Tradition should share the blame here. In the quest for damage stability improvement, design (passive) measures have traditionally been the only means to achieve it in a measurable/auditable way (SOLAS 2009, Ch. II-1). However, in principle, the consequences from inadequate damage stability can also be reduced by operational (active) measures, which may be very effective in reducing loss of life (the residual risk). There are two reasons for this. The first relates to the traditional understanding that operational measures safeguard against erosion of the design safety envelop (increase of residual risk over time). The second derives from lack of measurement and verification of the risk reduction potential of any active measures. In simple terms, what is needed is the means to account for risk reduction by operational measures as well as measures that may be taken during emergencies. Such risk reduction may then be considered alongside risk reduction deriving from design measures.

Therefore, new measures for risk reduction (operational and in emergencies) should be considered in addition to design measures. What needs to be demonstrated and justified is the level of risk reduction and a way to account for it, the latter by adopting a formal process and taking requisite steps to institutionalise it.

Efforts in this direction lead to setting up a large scale EC-funded research Project, FLOODING ACCIDENT RESPONSE (FLARE), [1], described briefly below, following a brief exploration of the flooding residual risk in large passenger ships.

3. RESIDUAL RISK IN LARGE PASSENGER SHIPS

The assessment of damage stability of passenger ships considering flooding accidents (collision, grounding and contact), according to the probabilistic framework, results in calculating three A-Indices (EMSA III, Task 4, 2015, [2]):

- A collision A-Index ACN, calculated according to SOLAS 2009
- An index for grounding and contact accidents resulting in bottom damages AGR,B and
- An index for grounding and contact accidents resulting in side damages, AGR,S.

The latter two are also calculated using a methodology analogous to SOLAS 2009 but with accident statistics pertaining to bottom and side grounding accidents, the latter collated over many EU projects over the past 23 decades.

Deriving from this, In the EMSA III Project [2], a procedure has been presented for the derivation of a single Attained Survivability Index accounting for flooding accidents (collision, grounding and contact), based on the "preservation of risk to human life", expressed in terms of the Potential Loss of Life (PLL) by a formula in the following form:

$$PLL_i = POB \cdot c_i \cdot (1 - A_i)$$

Where,

- POB is the number of persons on board (considering assumptions with respect to occupancy)
- i stands for flooding accidents (i.e., collision accidents – CN; bottom grounding and contact accidents – GR-B; side grounding and contact accidents – GR-S
- c_i is an appropriate coefficient, depending to the type of accident and the type of ship (RoPax or Cruise ship) that may be calculated from the corresponding risk model.

Assuming for the sake of argument a ship with 10,000 POB, a combined attained index of 0.9 and that society is prepared to pay roughly $8M to save a life in accordance with current IMO-adopted cost0effectivenesss criteria, then the Residual Risk is €8 billion. This is the residual risk, justifying that damage limitation must be a priority when it comes to megaships, also giving credence to the saying that IF YOU DOUBT THE VALUE OF SAFETY, TRY AN ACCIDENT!

4. EU PROJECT FLARE (FLOODING ACCIDENT RESPONSE)

In line with the above argument, the EU 2011 Transport White Paper and the Horizon 2020 transport programme aim to "reduce accident rates, fatalities and casualties", Project FLARE contributes to ongoing efforts to increase the safety of shipping by targeting the risk of flooding, which is of paramount importance for passenger ships. Taking into consideration specific procedures adopted by IMO for risk assessment and existing instruments (such as the Formal Safety Assessment), the main overriding objective of FLARE is to develop a Novel Risk-Based methodology beyond the existing state-of-the-art of 'live' flooding risk assessment and management in line with IMO high-level goals. Specific targets include:

- A risk aware life-cycle approach for effective flooding risk management, equally applicable to new and existing ships;
- The consideration of susceptibility to flooding due to collision, contact and grounding in order to better understand cause and effect of accidents;
- The study of the relationship between active and passive measures of risk mitigation and control on the one hand, and risk prevention measures on the other, in order to render risk-control measures post-accident most effective;
- The consideration of both "static" and "dynamic" risk reduction, mitigation or control measures;
- The development of methods to quantify the effect of all pertinent measures examined.
- The technical basis and a proposal for the revision of relevant IMO regulations towards a framework for life-cycle flooding risk management.

In line with the argument presented in the previous paragraph, it is noted with interest that whilst understanding susceptibility to flooding is being targeted as a key objective to enable cost-effective consideration of all measures of flooding risk reduction, migration and control, the main emphasis of this large-scale EU project is damage limitation post-accident rather than on preventing the incident in the first place.

The framework to achieve this derives from life-cycle considerations as explained next.

5. LIFE-CYCLE FLOODING RISK MANAGEMENT

5.1 PHASE 1: DESIGN PHASE (Figure 1)

Decisions made during the design phase, shape safety over the whole life cycle. In this respect, use of advanced tools and exploiting knowledge in all forms at the design stage is most effective and, hence, highly desirable. The derived knowledge could be used in further design iterations or indeed stored in a knowledge base for use over the life-cycle of the vessel, including emergencies. Key to life cycle risk management is the understanding that both the operational phase (monitoring and management of residual risk) as well as emergencies (emergency response) depend crucially on having identified during the design phase pertinent design vulnerabilities that lead to critical accident scenarios in pre-specified loading and environmental conditions and on the basis of this having taken appropriate measures to ensure:

(a) vulnerabilities have been eradicated and design measures taken to reduce these
(b) evaluation of residual risk, including KPIs for monitoring purposes during the operational phase
(c) preparedness in addressing all identified critical scenarios in case of an emergency (emergency response)

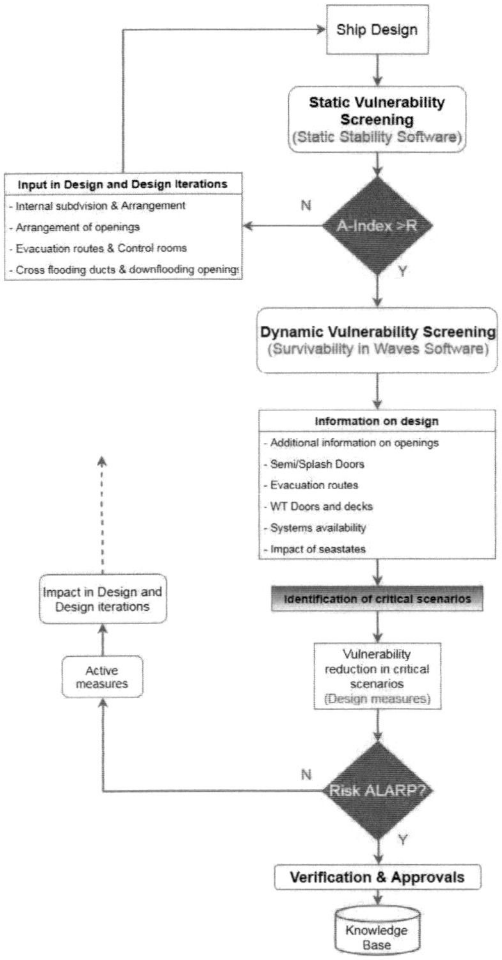

Figure 1: Design Phase

5.2 PHASE 2: OPERATIONAL PHASE (Figure 2)

Operation is the longest phase in the ship life-cycle, during which changes take place that have a direct impact on flooding risk. Therefore, risk must be monitored and reviewed to ensure changes in design / operation are reflected in the way risk is managed. This sounds straightforward in so far as changes take place in tangible, hence measurable, ship and environmental parameters, e.g., draft levels, loading condition, fluid tank levels, watertight door status, as well as prevailing wind and wave conditions. There are many other significant parameters and conditions, however, for example ship management, improving navigational equipment, training, safety culture and so on, which unarguably affect safety significantly. However, the impact of the majority of these factors on safety cannot, at present, be measured and hence monitored as there is not a system yet in place to assign risk credit/value to such influences. This will be one of the significant outcomes of EU Project SAFEMODE, [3].

Notwithstanding this, monitoring what can be measured and is known to be KPI for flooding risk in real time is a step in the right direction (this is depicted in Figure 2), to facilitate effective flooding risk monitoring, management and control.

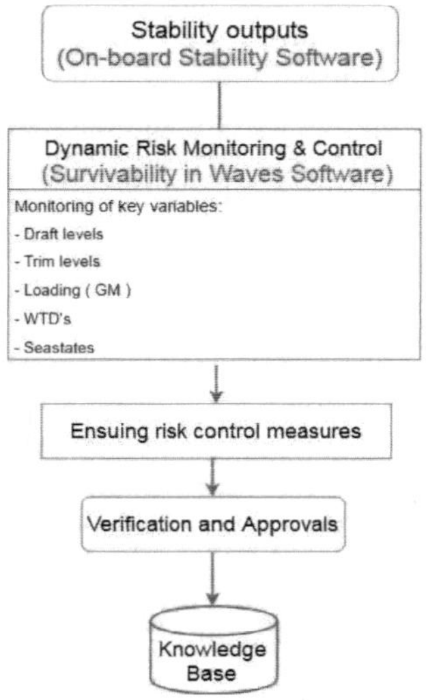

Figure 2: Operational phase

5.3 PHASE 3: EMERGENCY RESPONSE PHASE (Figure 3)

The emphasis, in this phase, is on ensuring that in any of the critical scenarios where ship stability is compromised, the ship will remain upright and afloat, with all safety systems available, for sufficient time to ensure safe evacuation of all the people on-board. However, similar to the operational phase most of the key elements affecting emergency response are not measurable/auditable, thus not providing any guidance to the operator in his strive to adopt cost-effective means to mitigate and control flooding risk in this vital last phase. Furthermore, new technologies are explored to enhance situational awareness and guidance of evacuees during emergencies, e.g., EC-funded Project SAFEPASS [4], interventional measures to restore stability post damage, [5] and so on. All of these, provide a platform for transformational changes in the way emergencies onboard are currently being addressed.

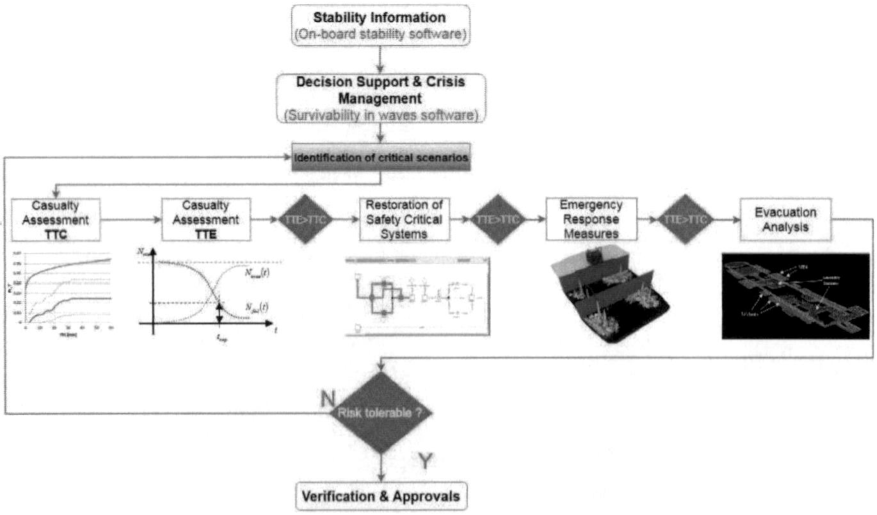

Figure 3: Emergency Response Phase

6. RECENT/ONGOING DEVELOPMENTS

A ship is designed to support and carry payload from place A to place B. Support is provided by buoyancy, which is dictated by the shape and internal volume of the ship hull and superstructure (permeable volume). However, when internal spaces of a ship are flooded as a result of collision or grounding accidents, permeable volume is lost and depending on the amount and distribution of this, a ship may sink (inadequate buoyancy) or capsize (inadequate stability). The traditional way of protection has been to subdivide the internal volume into smaller parts (watertight transverse/longitudinal subdivision) so that if these were damaged individually or collectively up to limits dictated by international/national regulations, the ship remains upright (or heels up to a given range) and afloat. However, there is a limit on how much floodwater the ship can support, which forms the basis of meeting applicable legislation but also a limit on in-build safety post-flooding accidents. This creates a serious problem: with society striving for higher safety relentlessly, a physical limit is reached where the contribution from the aforementioned subdivision arrangements have reached saturation. Therefore, as indicated in the foregoing, the industry struggles to attain the progressively higher damage stability standards demanded of newbuildings. Moreover, existing ships are traditionally exposed to weight accumulation higher up in the ship, which "eats" into the design margins, leading to unsustainable stability management over the life cycle of ships.

Based on the aforementioned ship fundamentals and accounting for operational experience and best practice, Maritime Safety Innovations Ltd, a spin out of the University of Strathclyde and MINOVA (a multi-national foam producer and supplier) have developed a series of innovations using in-built (passive) and interventional (active) measures, comprising the use of high expansion foam to restrict and contain the amount of floodwater entering the ship hull post flooding accidents under the trademark name AREST (Adaptable Reconfigurable Safety Technology) Systems. The deployable systems have been designed to also address the second largest risk contribution in the marine industry, namely fire, providing active fire protection as per SOLAS.

Feasibility studies on a number of cruise ships and RoPax demonstrate unprecedented improvements in damage stability in a highly cost-effective way, [4].

7. CONCLUSIONS

Based on the aforementioned research and development considerations, the following colluding remarks are drawn:

- Damage stability continues to be the main risk contributor for all types of ships. At the same time, escalation in size of passenger ships continues to attract the focus of attention with society (IMO) demanded progressively higher damage stability standards.
- Saturation in traditional means of achieving such standards, necessitates a broadening of traditional practices to account for novel ideas to fill this vacuum.
- Key in this development is a regulatory framework that accounts for and gives credit to all means of flooding risk reduction, mitigation and control.

- A large EC-funded research project (FLARE) is taking steps to develop a regulatory framework to facilitate this whilst a number of stakeholders in the maritime industry have already taken the initiative to support and embrace innovative solutions in their quest to ameliorate flooding risk.

8. ACKNOWLEDGEMENTS

The support by Scottish Enterprise in underpinning fundamental research in the use of high expandable foam for damage stability recovery, [5] is gratefully acknowledged. Our gratitude is also extended to cruise (RCCL) and RoPax (Color Line) operators for their support in this quest.

9. REFERENCES

1. FLARE: H2020-MG-2018, 814753 FLooding Accident Response.

2. EMSA/OP/10/2013: "A study assessing the acceptable and practicable risk level of passenger ships related to damage stability".

3. SAFEMODE (2019): "Strengthening synergies between Aviation and maritime in the area of human Factors" https://cordis.europa.eu/project/id/814961

4. Next generation of life Saving appliances and systems for saFE and swift evacuation operations on high capacity PASSenger ships in extreme scenarios and conditions 815146 — SafePASS — H2020-MG-2018-2019-2020/H2020-MG-2018-

5. Vassalos, D, Paterson, D and Boulougouris, E, "Water, Water Everywhere: Can High Expansion Foam Significantly Reduce Vessel Flooding Risk?", MT Journal, April 2019, pp42-49/

9. AUTHORS BIOGRAPHY

Dracos Vassalos is Professor of Maritime Safety at Strathclyde and Chairman of MSI Ltd, a University spinout offering innovative damage stability solutions. Chair of "Design for Safety" Conference, a topic he introduced in maritime industry and spearheaded its development with 100+ R&D projects. He is the founder of many maritime safety initiatives that helped shape safety research in academia and catalyse implementation in industry. Professor Vassalos received a Life Achievement Award from the Royal Academy of Engineering in 2011, the Gold Medal from the Royal Institution of Naval Architects in 2012, the Gold Medal from the Society of Naval Architects and Marine Engineers in the USA in 2015 and a Doctor of Science Degree from the University of Strathclyde in 2016. In 2019, he was elected Fellow of the Royal Academy of Engineering in the UK.

Donald Paterson is a PhD researcher at the Maritime Safety Research Centre of the University of Strathclyde, where he has studied and worked as a research assistant since completing his MEng degree in naval architecture and marine engineering in 2015. His research focuses on the wider subject of ship damage stability, with particular emphasis placed on the use of active/operational measures as an alternative means of damage stability enhancement. Since late 2018, he also has assumed the role of Naval Architect at Maritime Safety Innovations Ltd.

SOLAS 2020 DAMAGE STABILITY AND BEYOND

K. W. Hutchinson, Safinah Group, UK
A. L. Scott, Maritime and Coastguard Agency, UK

SUMMARY

Drawing upon the information and arguments put forward in papers read by the authors at the four previous RINA 'Damaged Ship' International Conferences, held in 2011 [1], 2013 [2], 2015 [3] and 2018 [4], and also at RINAs 'Design and Operation of Ferries and RO-PAX Vessels' International Conference of 2016 [5] and the recent joint HIMT and RINA International Conference on 'Sustainable and Safe Passenger Ships' [6], together with articles recently published in RINA's The Naval Architect magazine [7][8], this paper summarises all the amendments made to the SOLAS2009 Regulations and Explanatory Notes [9] in readiness for the SOLAS2020 version which entered into force on 1st January this year (IMO Resolutions MSC.421(98) [10] and MSC.436(99) [11]). It updates our paper on this subject to the 'Damaged Ship IV' Conference [4] in the light of further amendments to the SOLAS text and the accompanying Explanatory Notes agreed at SDC 6 in February 2019 and SDC 7 in February this year, most of which entered into force in 1st January 2020 but some of which are scheduled to enter into force on 1st January 2024.

Although the application dates of some of the Regulations and Explanatory Notes are now stretching beyond 2020, it is worthy of note that there is no agenda item currently scheduled on revisions to SOLAS Chapter II-1 or the Explanatory Notes for SDC 8 in January 2021 nor, for the first time in many years, will there be an inter-sessional Correspondence Group on Subdivision and Damage Stability, SDS CG. The chairman of the SDS Working Group at IMO, who was also the co-ordinator of the SDS CG, Mr James L. Person of the USCG, announced his retirement at SDC 7 after many years of excellent work in piloting representatives from Member States all over the world through the intricacies of changing SOLAS Chapter II-1 from a long-standing deterministically-based set of subdivision and damage stability regulations to ones based on probabilistic methods.

We have tried to keep RINA updated with these developments over the years but it does seem, at last, that the pace of changes to the SOLAS Chapter II-1 damage stability Regulations and Explanatory Notes is slowing down which will no doubt come as some relief to all those involved in the design, construction, approval and operation of Passenger and Dry Cargo ships.

NOMENCLATURE

A	Attained Subdivision Index
A_{max}	Maximum Attained Index
AP	Aft Perpendicular
bhd	Bulkhead
B	Breadth (metres)
BWM	Ballast Water Management
DB	Double Bottom
DC	Damage Control
EC	European Commission
EN	Explanatory Notes
ER	Engine Room
EU	European Union
GM_T	(Fluid) Transverse metacentric height (metres)
GZ	Righting lever (metres)
GZ_{max}	Maximum positive righting lever (metres) to θ_v. not to be taken as more than TGZ_{max} in relation to the calculation of $s_{final,i}$
HIMT	Hellenic Institute of Marine Technology
ICLL	International Convention on Load Lines
IE	Inclining Experiment
IMO	International Maritime Organization
IS	Intermediate Stage (of flooding)
KG_f	Fluid vertical centre of gravity (metres)
LLH	Long Lower Holds
L	Length (metres)
L_S	Subdivision length (metres)
LSA	Life Saving Appliances
MCA	UK Maritime and Coastguard Agency
MSC	IMO Maritime Safety Committee
N	Total number of PoB (SOLAS2020)
N	$N_1 + 2N_2$ (SOLAS2009)
N_1	Number of persons for whom lifeboats are provided (SOLAS2009)
N_2	Number of persons (including officers and crew) the ship is permitted to carry in excess of N_1 (SOLAS2009)
O/S	Open / Shut
p_i	Probability that only one compartment(s) under consideration is flooded
PAX	Passenger
PoB	Persons on Board
R	Required Subdivision Index
Range	Range of positive righting levers measured from θ_e. The positive range is taken up to θ_v (degrees) but not to be taken as more than *TRange* in relation to the calculation of $s_{final,i}$
RINA	The Royal Institution of Naval Architects
RO-PAX	Roll-On / Roll-Off Passenger
RO-RO	Roll-On / Roll-Off
RQD	Raised Quarterdeck

s_i (Conditional) Probability of survival of flooding following damage to the compartment(s) under consideration

$s_{intermediate,i}$ (Conditional) Probability of survival in the intermediate stages of flooding following damage to the compartment(s) under consideration

$s_{final,i}$ (Conditional) Probability of survival in the final stage of flooding following damage to the compartment(s) under consideration

SDC IMO Sub-Committee on Ship Design and Construction

SDS IMO SLF / SDC Subdivision and Damage Stability

SDS CG IMO SLF / SDC Subdivision and Damage Stability Correspondence Group

SIB Stability Information Booklet

SLF IMO Sub-Committee on Stability and Load Lines and on Fishing vessels (SDC from 2014)

SOLAS Safety of Life at Sea Convention. This paper focusses only on Chapter II-1 Parts B, B-1, B-2, and B4 (subdivision and stability)

SOLAS90 Version superseding SOLAS60 and the version prior to SOLAS2009 and applicable to ships constructed prior to 1st January 2009 – the last version based on deterministic damage stability assessment for Passenger ships, Dry Cargo ship damage stability requirements introduced based on a probabilistic assessment

SOLAS2009 Previous version to SOLAS2020 and is applicable to new ships constructed on or after 1st January 2009 – based on a probabilistic damage stability assessment for both Dry Cargo and Passenger ships

SOLAS2020 Current version which is applicable to new ships constructed on or after 1st January 2020 – based on probabilistic damage stability assessment for Dry Cargo and Passenger ships

SWTD Sliding Watertight Door

T Draught (metres)

TGZ_{max} Maximum GZ_{max} for ship/damage type (metres)

TRange Maximum *Range* for ship/damage type (degrees)

UK United Kingdom

US United States of America

USCG US Coast Guard

WL Waterline

WOD Water on Deck

WT Watertight

θ_e Equilibrium heel angle in any stage of flooding (degrees)

θ_{min} Minimum heel angle in any stage of flooding (degrees) in relation to the calculation of $s_{final,i}$

θ_{max} Maximum heel angle in any stage of flooding (degrees) in relation to the calculation of $s_{final,i}$

θ_v Angle, in any stage of flooding, where the righting lever becomes negative, or the angle at which an opening incapable of being closed weathertight becomes submerged (degrees)

1. INTRODUCTION

The basis text referred to in this paper is the SOLAS Consolidated Edition 2014 [9] which incorporates all SOLAS Amendments in effect from 1st July 2014, most notably from the damage stability aspect, the December 2006 Annex 2 amendments, adopted by IMO Resolution MSC.216(82) [12], which entered into force for Passenger ships and Dry Cargo ships with keels laid on or after 1st January 2009 [13].

Commonly known as the SOLAS2009 Amendments, MSC.216(82) [12] included revisions to, amongst other things, SOLAS Part 1 Chapter II-1 Parts A (General), B (Subdivision and Stability), B-1 (Subdivision and Damage Stability of Cargo Ships) and Part C Regulation 35-1 (Bilge Pumping Arrangements). The SOLAS2009 Amendments replaced the deterministic damage stability and subdivision regulations for Passenger ships in force prior to 1st January 2009 (commonly referred to as SOLAS90 [14]) with a new probabilistic damage stability calculation procedure, removing such concepts as floodable and permissible length, factor of subdivision, criterion of service and margin line many of which had been introduced in 1929 following the establishing of the first SOLAS Convention in 1914 as a result of the loss of the RMS Titanic in 1912.

For Dry Cargo ships of 80 metres in length and above, the previous probabilistic damage stability regulations in SOLAS90 Part B-1 Regulations 25-1 to 25-10 (IMO Resolution MSC.47(66) [15]) were harmonized with new probabilistic regulations for Passenger ships and all were accompanied by a set of Explanatory Notes (IMO Resolution MSC.281(85) [16]). Also replaced was IMO Resolution A.265(VIII) [17], which, since its introduction in 1973, had been considered as a probabilistic equivalent to all the deterministic SOLAS subdivision and damage stability regulations for Passenger ships in force between 1973 and 31 December 2008.

Since 1st January 2009, further amendments were found to be necessary to SOLAS2009. After considerable debate and discussion at IMO over the years these were finally approved under IMO Resolution MSC.421(98) [10], adopted on 15th June 2017, and IMO Resolution MSC.436(99) [11], adopted on 24th May 2018, for entry into force on 1st January 2020. They are now known as the SOLAS2020 Amendments, applicable to Passenger and Dry Cargo ships constructed on or after that date.

At the same time as these Amendments to the SOLAS2009 Regulations appeared, the first amendments to the accompanying Explanatory Notes were made and appeared in IMO Resolution MSC.429(98) [18] which was adopted on 9th June 2017, also for entry into force on 1st January 2020.

For ease of reference, the Amendments and Explanatory Notes relevant to each specific Regulation are shown in the Tables given in the Appendix to this paper, namely the 2020 amendments to SOLAS2009. It should be noted that Amendments to the text of SOLAS Chapter II-1 and updates to the Explanatory Notes continue to be made and will enter into force at dates now usually specified in the text. Hence, the Tables appended also detail these post-2020 SOLAS Amendments and Explanatory Notes agreed at SDC 6 (see SDC 6/13 [19] Annex 5 for details) and approved at MSC 101 in May 2019 together with those agreed at SDC 7 (see SDC 7/WP.4 [20]) for approval and adoption at MSC 102 in May 2020.

At SDC 7 in February 2020 further revisions to the 2020 Explanatory Notes were agreed (see the Appendix to this paper) and a consolidated version including all the amendments made to the Explanatory Notes since 2009 is expected to be approved and adopted at MSC 102 in May 2020 under IMO Resolution MSC.429(98)/REV.1, which will revoke MSC.429(98) [18]. Full details are in SDC 7/WP.4 [20] Annex 1. This working paper will eventually be incorporated as an annex to the SDC Report to approval at MSC 102.

Since April 2003 all EU flag RO-RO Passenger ships and non-EU flag ships operating in EU waters must additionally comply with the Stockholm Agreement (EU Directive 2003/25/EC, as amended [21]) which was implemented into UK law under SI 2004 No. 2884 [22] and MSN 1790(M) [23] (both as amended). It was originally intended that these requirements would be revoked on 1st January 2009 but concerns over the safety levels provided by the SOLAS2009 Amendments led to a decision by the EU to retain the Stockholm Agreement to be used in conjunction with the SOLAS2009 Amendments.

The subsequent Amendments to SOLAS2009 [10][11], which entered into force on 1st January 2020, have still not completely allayed the safety concerns. At the time of writing (March 2020), the Stockholm Agreement has not been revoked so must continue to be applied to all RO-RO Passenger ships constructed on or after 1st January 2009 including those to which the SOLAS2020 Amendments apply. A decision by the EU on the future of the Stockholm Agreement is awaited following the results from a three-year research project [24].

Now that the SOLAS2020 Amendments to SOLAS2009 are almost finalised at IMO, the Appendix to this paper briefly describes, in tabular form for easy reference, the changes made and some of the reasons for those changes, along with those in the Explanatory Notes. The paper gives a brief account of the main changes agreed at SDC 6 and SDC 7 then highlights one or two of the most significant amendments since 2009.

2. AMENDMENTS MADE AT SDC 6

The agenda item for SDC 6 was entitled 'Review SOLAS Chapter II-1, Parts B-2 to B-4, to ensure consistency with parts B and B-1 with regard to watertight integrity'. The basis for this agenda item was a paper to SDC 5 by Norway (SDC5/5 [25]) which pointed out that many of the provisions in Parts B-2 and B-4 in SOLAS90 were not within scope of the harmonization process leading to the SOLAS2009 Amendments. This resulted in a clash between the deterministic and probabilistic elements of the 2009 Regulations and a consequent lack of consistency. Norway pointed out by way of an example that Parts B-2 and B-4 still assume that the bulkhead deck can be used as a watertight boundary whereas this is no longer necessarily the case following the introduction of the probabilistic regulations where the old requirement for the bulkhead deck to be at least weathertight was removed. Norway's paper included several other examples of inconsistencies between deterministic Parts B-2 and B-4 and probabilistic Part B-1, and this attempt to resolve these issues formed the basis of much of the work at SDC 6 and subsequently. The following topics were discussed and resolved.

2.1 INCONSISTENCY BETWEEN PART B-1 REGULATION 7-2.5 AND PART B-2 REGULATION 17

Regulation 7-2.5 describes when the s_i factor is to be taken as zero following immersion of openings such as air pipes, ventilators and openings which are closed by weathertight doors or hatch covers, any of which can lead to progressive flooding. Whereas weathertight doors in locations above the bulkhead deck are assumed in this probabilistic regulation to be permitted for both Passenger and Cargo ships as long as the opening is accounted for in all damage cases, deterministic Regulation 17, which applies only to Passenger ships, requires that where openings are carried through watertight bulkheads or decks within the immersed part of the bulkhead deck:

... arrangements shall be made to ensure the watertight integrity of the structure above the bulkhead deck.

The difference is the assumption of watertightness in the deterministic regulations against the acceptance of non-watertightness in the probabilistic calculations as long as any openings through which progressive flooding may occur are accounted for in the damage calculations.

At SDC 6, this inconsistency was resolved by, in Regulation 7-2.5.2, making the existing provision in sub-Paragraph .1 applicable to Cargo ships only and then introducing a new sub-Paragraph .3 with exactly the same technical provision but applicable only to Passenger ships constructed before 1st January 2024. In Regulation 7-2.5.3, which refers to intermediate stages of

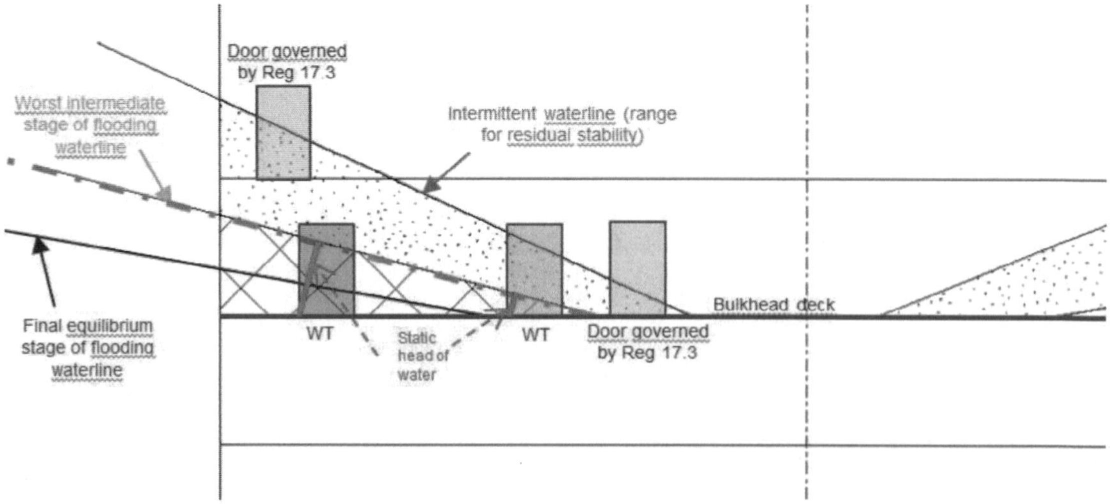

Figure 1: Illustrating the types of doors to be used following various flooding scenarios [MSC.429(98)/REV.1]

flooding, the same provision has been added in a new sub-Paragraph .4 for Passenger ships constructed on or after 1st January 2024.

In Regulation 17.1 the existing deterministic provision for watertightness has been retained but only for Passenger ships constructed before 1st January 2024. Ships constructed on or after 1st January 2024 must then comply with a probabilistic approach for accounting for the entry and spread of water along the bulkhead deck (new Regulation 17.2). Where pipes, scuppers, electric cables etc. are carried through watertight boundaries that are immersed at any intermediate or final stage of flooding in cases contributing to the Attained Index A, such openings shall be made watertight. However, doors in internal watertight subdivision arrangements above the bulkhead deck for Passenger ships constructed on or after 1st January 2024 will be covered by a new sub-Paragraph 17.3 which requires them to be capable of:

... preventing the passage of water when immersed in the required range of positive stability for any cases contributing to the attained subdivision index "A". These doors may remain open provided they can be remotely controlled from the navigation bridge. They shall always be ready to be immediately closed. (with the ship listed 15 degrees either way, according to the Explanatory Notes).

The term 'preventing the passage of water' is used to avoid the use of the terms:

'weathertight' meaning that water will not penetrate into the ship in any sea conditions; or

'watertight' meaning having scantlings and arrangements capable of preventing the passage of water in any direction under the head of water likely to occur in intact and damaged conditions. In the damaged condition, the head of water is to be considered in the worst situation at equilibrium, including intermediate stages of flooding;

neither of which are particularly suitable for the issue in question. The Explanatory Notes contain a new figure to illustrate the new regulation (see Figure below) and a new table (not included in this paper).

2.2 REGULATION 12.6.1 (FORMERLY 12.5.1) RELATING TO THE VALVES TO BE USED ON PIPES PENETRATING THE COLLISION BULKHEAD

The SOLAS2009 Amendments for Regulation 12.5.1 (now 12.6.1) made provision for screw-down valves operated from above the bulkhead deck with a valve chest fitted inside the fore peak to be used on pipes penetrating the collision bulkhead. Alternatively, the Administration:

... may, however, authorize the fitting of this valve on the after side of the collision bulkhead provided that the valve is readily accessible under all service conditions and the space in which it is located is not a cargo space.

Some Member States pointed out that many Cargo ships were in service which used butterfly valves and these were found to be safe and accepted in practice by surveyors, added to which replacements were much more readily available than for screw-down valves. A new sentence was added to the regulation to this effect:

Alternatively, for cargo ships, the pipe may be fitted with a butterfly valve suitably supported by a seat or flanges and capable of being operated from above the freeboard deck.

Later, following extensive further debate, it was decided at SDC 6 to restrict this amended Regulation to ships constructed on or before 1st January 2024 and introduce a more generalised new provision be made for ships constructed on or after 1st January 2024. Deletions are indicated below by ~~strikeout~~ and additions by underlining, with Explanatory Notes prefixed with **EN** in order to differentiate from the SOLAS text, as is the case throughout the body of the paper.

6.~~1~~2 For ships constructed on or after 1 January 2024, except as provided in paragraph 6.~~2~~3, the collision bulkhead may be pierced below the bulkhead deck of passenger ships and the freeboard deck of cargo ships by not more than one pipe for dealing with fluid in the forepeak tank, provided that the pipe is fitted with a ~~screw-down valve~~ remotely controlled valve capable of being operated from above the bulkhead deck of passenger ships and the freeboard deck of cargo ships., ~~the valve being located inside the forepeak at the collision bulkhead. The Administration may, however, authorize the fitting of this valve on the after side of the collision bulkhead provided that the valve is readily accessible under all service conditions and the space in which it is located is not a cargo space. Alternatively, for cargo ships, the pipe may be fitted with a butterfly valve suitably supported by a seat or flanges and capable of being operated from above the freeboard deck.~~ The valve shall be normally closed. If the remote-control system should fail during operation of the valve, the valve shall close automatically or be capable of being closed manually from a position above the bulkhead deck of passenger ships and the freeboard deck of cargo ships. The valve shall be located at the collision bulkhead on either the forward or aft side, provided the space on the aft side is not a cargo space. ~~All~~ The valves shall be of steel, bronze or other approved ductile material. Valves of ordinary cast iron or similar material are not acceptable.

The Explanatory Notes now include illustrations of acceptable configurations of butterfly valve for ships constructed before 1st January 2024, as shown in Figure 2.

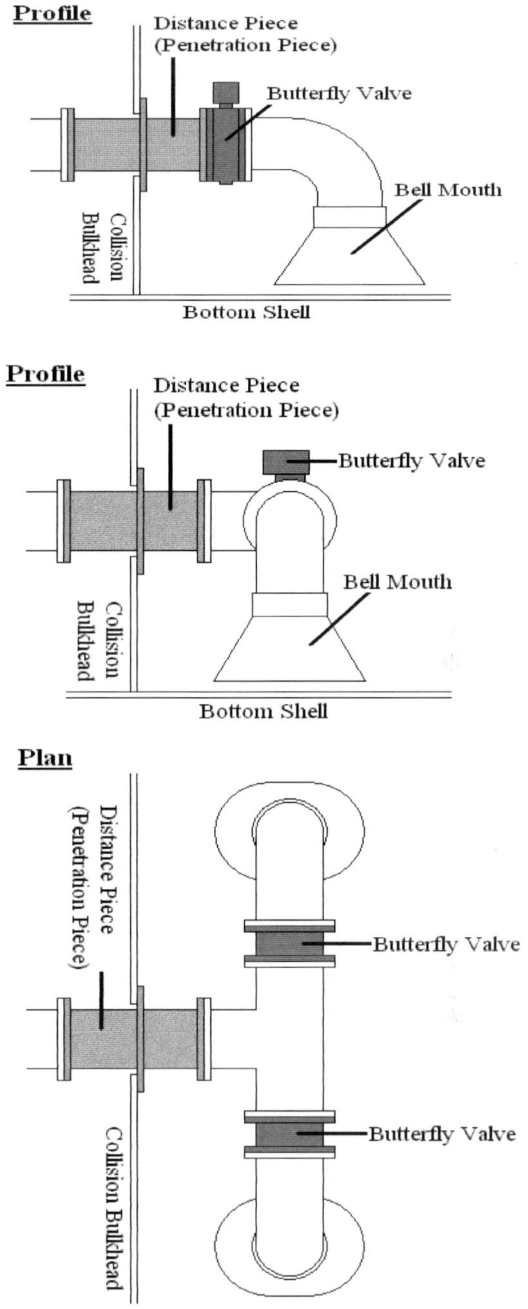

Figure 2: Illustrating acceptable butterfly valve arrangements on pipes penetrating the collision bulkhead [MSC.429(98)/REV.1]

2.3 AMENDMENTS TO REGULATION 13 RELATING TO OPENINGS IN WATERTIGHT BULKHEADS BELOW THE BULKHEAD DECK IN PASSENGER SHIPS

The main focus of attention with this Regulation was to clarify and, to some degree, account for well-established existing practice with regard to indication as to the status of closure of power-operated sliding watertight doors by visual and audible alarms and where such indication should be located. SOLAS2009 limited the location to the navigation bridge and at all locations where hand-operation of the doors takes place above the bulkhead deck. It was pointed out that as SOLAS Chapter 11-2 Regulation 23 now requires all Passenger ships constructed after 1 July 2010 to be equipped with a safety centre it would be logical to make this the location of a central operating console for all power-operated sliding watertight doors and so the existing provisions in former Regulation 13.8.1 (now 13.7.1) were modified accordingly:

~~8~~7.1 A central operating console for all power-operated sliding watertight doors shall be located in the safety centre in accordance with regulation II-2/23. If the safety centre is located in a separate space adjacent to the navigation bridge, a central operating console shall also be located on the navigation bridge. The central operating console(s) ~~at the navigation bridge~~ shall have a "master mode" switch with two modes of control: a "local control" mode which shall allow any door to be locally opened and locally closed after use without automatic closure, and a "doors closed" mode which shall automatically close any door that is open in not more than 60 s with the ship in an upright position. [~~The "doors closed" mode shall automatically close any door that is open.~~] The "doors closed" mode shall permit doors to be opened locally and shall automatically re-close the doors upon release of the local control mechanism. The "master mode" switch shall normally be in the "local control" mode. The "doors closed" mode shall only be used in an emergency or for testing purposes. ~~Special consideration shall be given to the reliability of the "master mode" switch.~~

Regulation 13.7.2 (formerly 13.8.2) which includes the term 'navigation bridge' is now applicable only to Passenger ships constructed before 1st January 2024. For ships constructed on or after 1st January 2024 this same provision applies under Regulation 13.7.3 but with the reference to 'navigation bridge' removed and a clarification that the door diagram should only show each power-operated sliding watertight door rather than 'each door'. In other words, the diagram should focus attention on the safety-critical doors to avoid information overload. The regulation also requires the door location and open/closed status to be shown on the onboard Stability Instrument / computer, if installed in accordance with Regulation II-1/8-1.3.1.

2.4 REGULATION 15 RELATING TO OPENINGS IN THE SHELL PLATING BELOW THE BULKHEAD DECK OF PASSENGER SHIPS AND FREEBOARD DECK OF CARGO SHIPS

Existing Regulation 15.9 relating to the watertightness of gangway, cargo and fuelling ports in a ships' side below the bulkhead / freeboard deck will now only apply to ships constructed before 1st January 2024. For ships constructed after this date the regulation has been modified as below:

~~9~~10 For ships constructed on or after 1 January 2024, ~~Gangway, c~~Cargo ports and other similar openings (e.g. gangway and fuelling ports) in the side of ships ~~fitted~~ below the bulkhead deck of passenger ships and the freeboard deck of cargo ships shall be fitted with doors so designed as to ensure the same watertightness and structural integrity as the surrounding shell plating. Unless otherwise granted by the Administration, these openings shall open outwards. The number of such openings shall be the minimum compatible with the design and proper working of the ship. ~~watertight and I~~ In no case shall these openings be so fitted as to have their lowest point below the deepest subdivision draught.

Existing Regulation 10.1 relating to ash chutes and rubbish chutes was regarded as obsolete and therefore deleted. Regulation 10.2 was also deleted as it related to 10.1 and made reference to watertight covers and automatic non-return valves.

2.5 REGULATION 16 RELATING TO THE CONSTRUCTION AND INITIAL TESTS OF WATERTIGHT CLOSURES

The word 'closures' replaced 'watertight decks, trunks, etc.' in the title of this Regulation 16 to make it more general in scope. As such, it was realised that hatches had not previously been included in these regulations as being required to be watertight so are now included but references to ash chutes and rubbish chutes have been deleted as for Regulation 15.

2.6 REGULATION 17-1 RELATING TO THE INTEGRITY IF THE HULL AND SUPERSTRUCTURE, DAMAGE PREVENTION AND CONTROL ON RO-RO PASSENGER SHIPS

The wording of Paragraphs 1.1 to 1.3 has been amended and the Explanatory Notes (**EN**) revised, as detailed below:

1.1 ~~Subject to the provisions of paragraphs 1.2 and 1.3, a~~All accesses from the ro-ro deck that lead to spaces below the bulkhead deck shall have a lowest point which is not less than 2.5 m above the bulkhead deck, unless the access is covered by the provisions in paragraphs 1.2 or 1.3.

1.2 Where vehicle ramps are installed to give access to spaces below the bulkhead deck, their openings shall be able to be closed weathertight to prevent ingress of water below, ~~alarmed and indicated to the navigation bridge~~ and fitted with alarms and open/close indicators on the navigation bridge. The means of closure shall be watertight if the deck is intended as a watertight horizontal boundary under regulation 7-2.6

EN *Regulation 17-1.1.2* *If a non-watertight vehicle ramp closure is assumed to restrict the flow of water during the calculation of the attained subdivision index A, the vehicle ramp opening should comply with regulation 7-2.5.3.4.*

1.3 Subject to regulations 23.3 and 23.6, t~~T~~he Administration may permit the fitting of particular accesses to spaces below the bulkhead deck provided they are necessary for the essential working of the ship, e.g. the movement of machinery and stores, and subject

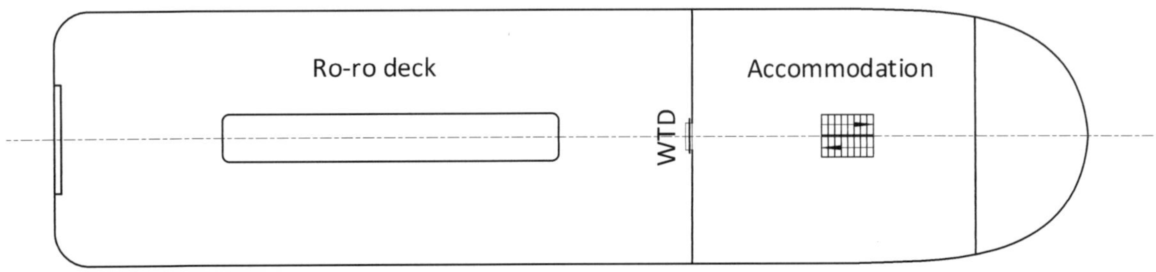

Figure 3: Illustration of horizontal access from a RO-RO space through a watertight door [MSC.429(98)/REV.1]

to such accesses being made watertight, ~~alarmed and indicated on the navigation bridge~~ and fitted with alarms and open/close indicators on the navigation bridge.

EN *Regulations 17-1.1.1 and 17-1.1.3* *apply only to direct accesses from a ro-ro space to spaces located below the bulkhead deck. The operation of watertight doors in bulkheads separating a ro-ro space and other spaces as per regulation 13.8.1 should be limited to compliance with regulation 23.3.*

Of particular interest is the new provision in Paragraph 17-1.1.2 stipulating that the means of closure of a vehicle ramp shall be watertight if the bulkhead deck is intended as a watertight horizontal boundary under Regulation 7-2.6. The latter regulation states that:

Where horizontal watertight boundaries are fitted above the waterline under consideration the s-value calculated for the lower compartment or group of compartments shall be obtained by multiplying the value as determined in paragraph 1.1 by the reduction factor v_m according to paragraph 6.1, which represents the probability that the spaces above the horizontal subdivision will not be flooded.

Explanatory Note 17-1.1.2 confirms that if a non-watertight vehicle ramp closure is assumed to restrict the flow of water for the purposes of calculating index *A* it will, after 1st January 2024 be subject to new Regulation 7-2.5.3.4, which states:

The factor s_i is to be taken as zero if, taking into account sinkage, heel and trim, any of the following occur in any intermediate stage or in the final stage of flooding:
.4 for passenger ships constructed on or after 1 January 2024, immersion of the lower edge of openings through which progressive flooding may take place and such flooding is not accounted for in the calculation of factor si. Such openings shall include air-pipes, ventilators and openings which are closed by means of weathertight doors or hatch covers.

In other words, immersion of the non-watertight vehicle ramp closure would result in the factor s_i being taken as zero and therefore the damage cases in question would make no contribution to *A*. The likelihood of a vehicle ramp closure being watertight at all times (how and when is this tested is yet to be established) would seem to be small and so these new Regulations and Explanatory Notes will need to be carefully applied by the Approving Authorities.

2.7 REGULATION 19 RELATING TO THE PROVISION OF DAMAGE CONTROL INFORMATION

A new Paragraph, 5, has been added requiring Passenger ships constructed on or after 1st January 2024 and subject to the Safe Return to Port Regulation 8-1.3 the damage control information should make reference to the activation of damage stability support from the onboard Stability Instrument / computer, if fitted or shore-based support when provided.

2.8 REGULATION 22 RELATING TO THE PREVENTION AND CONTROL OF WATER INGRESS, ETC.

Regulation 22.6 relating to the closure and making watertight of gangway, cargo and fuelling ports before the voyage commences has been amended and included as a new Regulation 22.7 as follows for ships constructed after 1st January 2024 with a new Explanatory Note added:

7 For ships constructed on or after 1 January 2024, ~~G~~gangway, cargo and fuelling ports fitted below the bulkhead deck of passenger ships and the freeboard deck of cargo ships and all watertight hatches, shall be effectively closed and secured watertight before the voyage commences, and shall be kept closed during navigation. However, the master may permit a watertight hatch to be opened during navigation for a limited period of time sufficient to permit passage or for access. It shall then be closed.

EN *Regulation* *22.7 This provision applies to any hatches that are considered watertight in the damage stability calculations, whether fitted above or below the bulkhead deck of passenger ships or the freeboard deck of cargo ships.*

This is to make provision for watertight hatches, which had hitherto been omitted from this paragraph, primarily for consistency with revised regulation 16.2 and for clarification of when they may be opened during navigation.

3. AMENDMENTS MADE AT SDC 7

The agenda item for SDC 7 was entitled 'Amendments to the Explanatory Notes to SOLAS chapter II-1 subdivision and damage stability regulations (Resolution MSC.439(98)'. The outcome of the discussions at SDC 7 was a revision of the Explanatory Notes in MSC.429(98) [18] into a new consolidated version to be known as MSC.429(98)/REV.1. This is currently to be found in paper SDC 7/WP.4 [20] but will be incorporated into an annex to the Report of the SDC to MSC 102 for approval in May 2020. It is therefore possible that some of the amendments may be changed or omitted at this time. The principle changes agreed at SDC 7 were as follows.

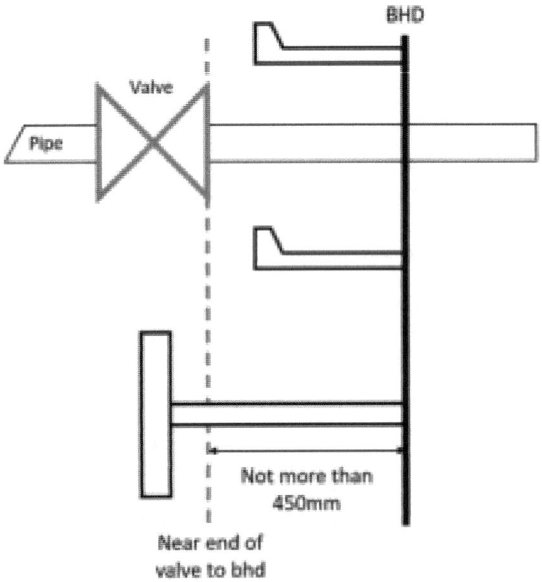

Figure 4: Illustrating the maximum separation distance of a valve from a bulkhead [MSC.429(98)/REV.1]

3.1 EXPLANATORY NOTE FOR REGULATION 7.7 RELATING TO THE PREVENTION OF PROGRESSIVE FLOODING THROUGH PIPES AND VALVES SITUATED CLOSE TO A BULKHEAD OR DECK

The original Explanatory Note had been revised to say that pipes and valves 'directly adjacent or situated as close as practicable' to a bulkhead or deck can be regarded as part of that bulkhead or deck when damaged and that the separation distance on either side of the bulkhead or deck is of the same order as the bulkhead or deck stiffening structure. This rather vague wording had been giving rise to some arguments within the industry and so a proposal to quantify the separation distance either side of the bulkhead or deck was proposed and accepted at SDC 7 for ships constructed after 1st January 2024. A new diagram was also included for further clarification:

EN *2. This explanatory note only applies to ships constructed on or after 1 January 2024. Pipes and valves directly adjacent or situated as close as practicable to a bulkhead or to a deck can be considered to be part of the bulkhead or deck, provided the separation distance on either side of the bulkhead or deck is of the same order as the bulkhead or deck stiffening structure. The same applies for small recesses, drain wells, etc. In no case should the separation distance on either side of the bulkhead or deck be more than 450 mm measured from the valve's near end to the bulkhead or deck (see figure below).*

Very similar amendments were made to the Explanatory Notes for Regulations 7-1.1.1.7-8 and 7-1.1.2.12-13

3.2 EXPLANATORY NOTE FOR REGULATION 22.3 REGARDING THE GUIDANCE ON WATERTIGHT DOORS WHICH MAY BE OPENED DURING NAVIGATION

This Explanatory Note clarifies that there is no prescribed guidance with respect to stability survivability in the event of watertight doors being left open on Cargo ships.

3.3 EXPLANATORY NOTE FOR REGULATION 23.6 CLARIFYING THE TERM 'ESSENTIAL WORKING OF THE SHIP' ON RO-RO PASSENGER SHIPS

This Explanatory Note clarifies that the movement of cargo during navigation is not considered as essential in terms of permitting some accesses to be opened during the voyage.

3.4 EXPLANATORY NOTES AGREED AT SDC 7 BUT ALREADY MENTIONED UNDER SECTION 2, ABOVE

Some of the Explanatory Notes, ENs, discussed and agreed at SDC 7 have already been mentioned adjacent to the regulatory changes covered in Section 2 of this paper and agreed at SDC 6. These include:

- 2.1 ENs for Regulations 7-2.5.2.1 and .3, 17.1, 17.2 and 17.3 relate to the resolution of the inconsistency between Part B-1 Regulation 7-2.5 and Part B-2 Regulation 17;
- 2.6 EN for Regulation 17-1.1.2 relating to non-watertight vehicle ramps;
- 2.8 EN for Regulation 22.7 relating to the inclusion of watertight hatches being closed before the voyage commences.

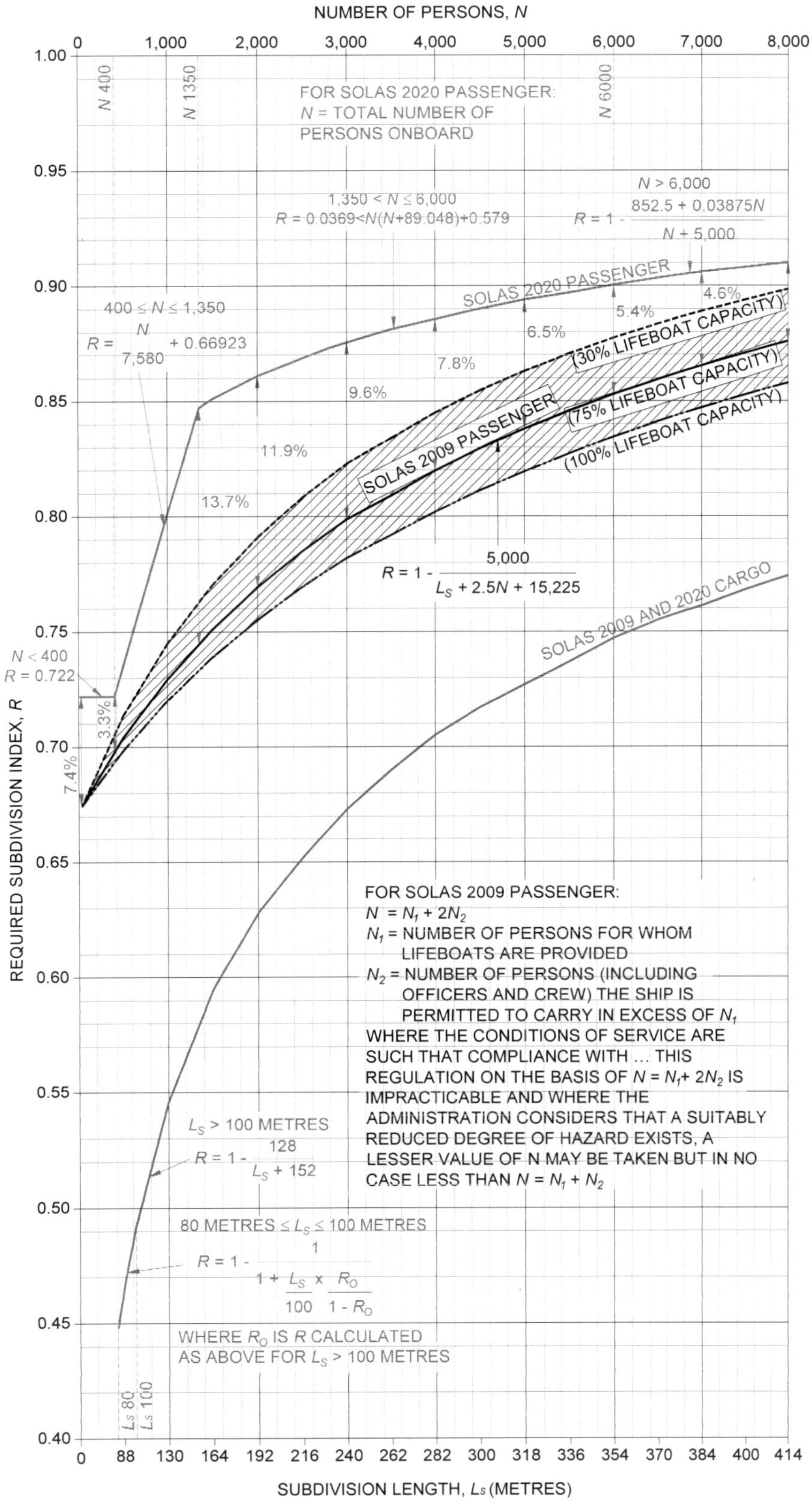

Figure 5: Plots of *R* against *Ls* and *N* to illustrate changes to SOLAS 2009 and 2020

3.5 SUMMARY

This completes the summary of work carried out on the revisions to SOLAS2009 Chapter II-1 and the accompanying Explanatory Notes at SDC 6 and SDC 7. A Table is included in an Appendix at the end of this paper for easy reference purposes. The resulting 2020 Amendments are already under review, noting, in particular, a new Regulation 25-1 relating to the provision of water-level detectors in the holds on multi-hold cargo ships which will enter into force on 1st January 2024.

4. CONCLUSIONS

The project to change the SOLAS damage stability regulations from a deterministic approach, in use for almost a century, to a much more complex but thorough probabilistic approach has been completed. An immense amount of work and effort has gone into checking and refining the resulting SOLAS2009 Regulations based on experience gained in using them to design and operate new ships. It is to be hoped that this will result in a new generation of even safer Passenger and Dry Cargo ships but that the 'existing' ships built between 1st January 2009 and 31st December 2019 will live out their careers safely transporting cargo and passengers along the world's seaways.

Perhaps the most significant amendment to SOLAS2009 was to uplift the R index as shown in the Figure 5 previously.

It can be seen from the above that the percentage increase in *R*, particularly for intermediate size RO-RO Passenger ships, is considerable and will pose a challenge to designers, especially for ships fitted with Long Lower Holds (LLH). In this respect, for European operators, the decision on whether the EU will decide to continue to apply the Stockholm Agreement is crucial.

Finally, was the change from deterministic to probabilistic damage stability worthwhile? Despite some of the difficulties encountered and complexities involved the answer is a qualified 'yes'. It is absolutely certain that the new Regulations could not be used for design purposes or by the approval authorities to check the designs without the use of ever more powerful computers and sophisticated software. This may be a cause for concern if too much reliance is placed on computers checking the results from other computers with little or no human intervention thereby losing that quality of having a 'hunch' that something is not quite right. Let us hope that the software engineers have it all under control! At least the pace of change appears to be slowing, giving everybody involved time to assimilate all the changes which have occurred since the SOLAS2009 Amendments came into force.

5. DISCLAIMER

The views expressed in this paper are those of the authors and do not necessarily represent those of the organisations with which they are affiliated and the professional institutions of which they are members.

6. REFERENCES

1. HUTCHINSON, K.W., SCOTT, A.L., WRIGHT, P.N.H., WOODWARD, M.D. and DOWNES, J., 'Consideration of Damage to Ships from Conceptual Design to Operation: The Implications of Recent and Potential Future Regulations regarding Application, Impact and Education', *Proceedings of the International Conference on The Damaged Ship*, The Royal Institution of Naval Architects, London, United Kingdom, 26th and 27th January, 2011. ISBN: 978-1-905040-79-7

2. HUTCHINSON, K.W. and SCOTT, A.L., 'The Development and Impact of Current and possible Future International Damage Stability Regulations', *Proceedings of the International Conference on The Damaged Ship II*, The Royal Institution of Naval Architects, London, United Kingdom, 30th and 31st January, 2013. ISBN: 978-1-909024-11-3

3. HUTCHINSON, K.W. and SCOTT, A.L., 'Current and possible future Intact and Damage Stability Passenger Ship Regulations, specifically the provision of Damage Stability Information and Verification Tools for the Master', *Proceedings of the International Conference on The Damaged Ship III*, The Royal Institution of Naval Architects, London, United Kingdom, 25th and 26th March, 2015. ISBN: 978-1-909024-38-0

4. HUTCHINSON, K.W. and SCOTT, A.L., 'SOLAS 2020 Passenger and Cargo Ship Damage Stability Regulations', *Proceedings of the International Conference on The Damaged Ship IV*, The Royal Institution of Naval Architects, London, United Kingdom, 16th and 17th May, 2018. ISBN: 978-1-909024-75-5

5. HUTCHINSON, K.W. and SCOTT, A.L., 'Passenger RO-RO Ferry Damage Stability: Status and Development of International Regulations', *Proceedings of the International Conference on The Design and Operation of Ferries and Ro-Pax Vessels*, The Royal Institution of Naval Architects, London, United Kingdom, 25th and 26th May, 2016. ISBN: 978-1-909024-54-0

6. HUTCHINSON, K.W. and SCOTT, A.L., 'Stability of RO-PAX Ships in European Waters: SOLAS 2020 and the Stockholm Agreement', *Proceedings of the International Conference on Sustainable and Safe Passenger Ships*, The Hellenic Institute of Marine Technology and The Royal Institution of Naval Architects, Athens, Greece, 4th March 2020. ISBN: 978-1-911649-04-5

7. HUTCHINSON, K.W. 'Damage stability update', The Naval Architect, The Royal Institution of Naval Architects, London, United Kingdom, May 2017. ISSN: 0306-0209

8. HUTCHINSON, K.W. and SCOTT, A.L., 'Damage stability: upcoming SOLAS 2020 regulations', The Naval Architect, The Royal Institution of Naval Architects, London, United Kingdom, September 2017. ISSN: 0306-0209

9. SOLAS Consolidated Edition, 2014, *Consolidated text of the International Convention for Safety of Life at Sea, 1974, and its Protocol of 1988: articles, annexes and certificates. Incorporating all amendments in effect from 1 July 2014,* International Maritime Organization, London, United Kingdom, 2014. ISBN 978-92-801-1594-9

10. Resolution MSC.421(98) *Amendments to the International Convention for the Safety of Life at Sea, 1974, as amended*, Maritime Safety Committee, International Maritime Organisation, London, United Kingdom, 15th June, 2017.

11. Resolution MSC.436(99) *Amendments to the International Convention for the Safety of Life at Sea, 1974, as amended*, Maritime Safety Committee, International Maritime Organisation, London, United Kingdom, 24th May, 2018.

12. Resolution MSC.216(82) *Amendments to the International Convention for the Safety of Life at Sea, 1974, as amended*, Maritime Safety Committee, International Maritime Organisation, London, United Kingdom, 8th December, 2006.

13. SOLAS Consolidated Edition, 2009, *Consolidated text of the International Convention for Safety of Life at Sea, 1974, and its Protocol of 1988: articles, annexes and certificates. Incorporating all amendments in effect from 1 July 2009,* International Maritime Organization, London, United Kingdom, 2009. ISBN 978-92-801-1505-5 (note Erratum for Part B-1: Stability Regulation 7-2 (3) on page 47 re $s_{final,i}$)

14. SOLAS Consolidated Edition, 1992, *Consolidated text of the International Convention for Safety of Life at Sea, 1974, and its Protocol of 1988: articles, annexes and certificates. Incorporating all amendments up to and including the 1990 amendments (1991 amendments as an appendix),* International Maritime Organization, London, United Kingdom, 1992. ISBN 978-92-801-1594-5

15. Resolution MSC.47(66) *Adoption of Amendments to the International Convention for the Safety of Life at Sea, 1974*, Maritime Safety Committee, International Maritime Organisation, London, United Kingdom, 4th June, 1996.

16. Resolution MSC.281(85) *Explanatory Notes to the SOLAS Chapter II-1 Subdivision and Damage Stability Regulation*, Maritime Safety Committee, International Maritime Organisation, London, United Kingdom, 4th December, 2008.

17. Resolution A.265(VIII) *Regulations on Subdivision and Stability of Passenger Ships as an Equivalent of Part B of Chapter II of The International Convention for the Safety of Life at Sea, 1960*, Assembly: Eighth Session, 13th to 23rd November 1973, Resolutions and Other Decisions. Inter-Governmental Maritime Consultative Organization, London, United Kingdom, 1974.

18. Resolution MSC.429(98) *Revised Explanatory Notes to the SOLAS Chapter II-1 Subdivision and Damage Stability Regulation*, Maritime Safety Committee, International Maritime Organisation, London, United Kingdom, 9th June, 2017.

19. SDC 6/13 *Report to the Maritime Safety Committee*, Sub-Committee on Ship Design and Construction, International Maritime Organisation, London, United Kingdom, 28th February, 2019.

20. SDC 7/WP.4 *Report of the Working Group on Subdivision and Damage Stability*, Sub-Committee on Ship Design and Construction, International Maritime Organisation, London, United Kingdom, UNPUBLISHED.

21. Directive 2003/25/EC of the European Parliament and of the Council of 14 April 2003 on specific stability requirements for ro-ro passenger ships, Official Journal L123, of the European Union, Brussels, Belgium, 17th May, 2003.

22. Statutory Instrument 2004 No.2884 *Merchant Shipping (Ro-Ro Passenger ships) (Stability) Regulations 2004*, The Stationary Office, London, United Kingdom, 29th November, 2004. ISBN 0-11-050076-8

23. Merchant Shipping Notice 1790(M) *Agreement Concerning Specific Stability Requirements for Ro-Ro Passenger Ships Undertaking Regular Scheduled International or Domestic Voyages between European Ports*, The Stationary Office, London, United Kingdom, January, 2005

24. *Assessment of specific EU stability requirements for ro-ro passenger ships*, Directorate-General for Mobility and Transport - European Commission, Luxembourg, January 2019. (DNVGL, 24th January, 2019) ISBN 978-92-76-02100-1

25. SDC 5/5 *Review SOLAS Chapter II-1, Parts B-2 to B-4, to ensure consistency with parts B and B-1 with regard to watertight integrity – Proposal for discussion and possible amendments*, Sub-Committee on Ship Design and Construction, International Maritime Organisation, London, United Kingdom, 19th October, 2017.

7. AUTHORS BIOGRAPHIES

Keith Hutchinson BEng(Hons) CEng CMarEng FRINA FIMarEST FSNAME is a Senior Consultant in Whole Ship Design and Naval Architecture with Safinah Group, where he undertakes commercial and offshore marine consultancy for a range of clients globally. He is a graduate in Naval Architecture and Shipbuilding from both Sunderland Polytechnic and Newcastle University. Having worked in both Shipbuilding and Shiprepair yards on the River Tyne, he joined Swan Hunter Shipbuilders in 1990 as a Naval Architect and then in 1995 Armstrong Technology, which was acquired by Babcock in 2000, becoming a Senior then the Principal Naval Architect and finally overall Engineering Manager for the Tyneside Ship Design office until its closure in 2018 - he was also Chief Engineer for Commercial Marine and Subject Matter Expert for the Naval Architecture domain across Babcock. Keith has published over fifty technical papers and articles and been invited to give numerous public presentations on ship and offshore vessel design and performance. He sits on a variety of professional technical committees developing standards and guidance for the maritime industry globally and is active in several professional learned institutions and their branches.

Andrew Scott MBE BSc CEng FRINA is Policy Lead for Stability, Load Line and Tonnage within the Ship Standards Division of the Directorate of Maritime Safety and Standards at the Maritime and Coastguard Agency (MCA). He is based in their Southampton headquarters, United Kingdom, but now works mostly from his home in the North East of England. He is responsible for policy matters relating to stability, load line and tonnage and was until recently co-ordinator of the International Maritime Organization's (IMO) Ship Design and Construction (SDC) Correspondence Group covering the latest updates to SOLAS 2009 and on issues relating to Roll-On / Roll-Off (RO-RO) stability. He has led several MCA-sponsored Research Programmes into the impact of the 2009 SOLAS amendments on ship design most specifically on the issue of Water-on-Deck (Stockholm Agreement). Andrew graduated from the University of Newcastle upon Tyne in 1969 in Naval Architecture and Shipbuilding and joined Lloyds Register of Shipping (LRS). In 1979 he moved to the British Ship Research Association (BSRA), later BMT, in Wallsend in 1979 as a Research Officer working in consultancy, particularly RO-PAX ships, before joining the MCA 2003.

APPENDIX – TABLES SUMMARISING AMENDMENTS MADE TO THE SOLAS2009 REGULATIONS AND EXPLANATORY NOTES

For ease of reference, the 2020 Amendments to the SOLAS2009 text are in normal black text with the 2020 amendments to the Explanatory Notes are in in *black italic*.

Amendments to SOLAS Chapter II-1 and the Explanatory Notes continue to be made and will enter into force at dates now usually specified in the text. In this Appendix these post-2020 SOLAS text Amendments are in blue text (see SDC 6/13 [19] Annex 5 for details) and were approved at MSC 101 in May 2019.

At SDC 7 in February 2020 further revisions to the 2020 Explanatory Notes were agreed (shown in *green italic*) and a consolidated version including all the amendments made to the Explanatory Notes since 2009 is expected to be approved and adopted at MSC 102 in May 2020 under Resolution MSC.429(98)/REV.1, which revokes MSC.429(98) [18]. Full details are in SDC 7/WP.4 [20] Annex 1. This working paper will eventually be incorporated as an annex to the SDC Report for approval at MSC 102.

Further Amendments to the SOLAS text were agreed at SDC 7 for approval and adoption at MSC 102 in May 2020 are shown in red text in this Appendix (see SDC 7/WP.4 [20]). These new Regulations will enter into force for Cargo ships constructed after 1st January 2024.

The main amendments to SOLAS2009 Chapter II-1 Parts A, B, B-1, B-2, B-4 and C (MSC 98/23/Add.1, Annex 3; Resolution MSC.421(98) [10] and MSC 99/22/Add.1 Annex 1; Resolution MSC.436(99) [11]) and the *Explanatory Notes* (EN) (MSC.429(98)/REV.1) contained at present in SDC 7/WP.4 [20], prior to approval at MSC 102 in May 2020, are shown in the following Appendix Tables.

PART A - GENERAL

Regulation	Subject	Amendment	Reason
A 1.1.1	Application of the SOLAS2020 revisions	In general, the text revisions only apply to ships constructed after 01/01/2020	There are significant changes made to SOLAS2009 by the 2020 revs
A 1.1.2.1	Application of SOLAS2009 Amendments	Defines the regulations applicable to ships constructed 01/01/2009 to 31/12/2019	Amendments and these new regs clarify the applicability dates
A 1.1.2.2	Application of Reg. 8-1.3.	PAX ships built pre 01/01/2014 to have computers by 1st survey after 01/01/2025	All PAX ships eventually to be fitted with onboard computers
A 1.1.2.2	Application of new Reg. 19-1	Applies new Reg. 19-1 on DC drills to all PAX ships, post 01/01/2009	Damage control drills required on all PAX ships, post *Costa Concordia*
A.1.1.3.2	Application of the SOLAS2024 revisions	Defines the regulations applicable to ships constructed after 01/01/2024	Changes to SOLAS2020 entering into force 01/01/2024
A 1.1.3.4	Cargo ship alterations	Deleted	A/R ratio not relevant for new SOLAS2009 ships
A 1.2.2	Application of new Reg. 19-1	Applies new Reg. 19-1 on DC drills to all PAX ships, pre 01/01/2009	Damage control drills required on all PAX ships, post *Costa Concordia*
A 1.3	*Application of regs after major alterations*	*Clarifies what to do with A for ships built 01/01/2009 - 31/12/2019 and altered*	*Response to questions from industry for clarification of SOLAS text*
A 2.2 2 .9 2.10 2.19	Definitions	Draught, trim, bulkhead deck definitions modified	Increased clarity
A.2.11	*Light service draught*	*Clarifies differences between Cargo and PAX ships also with ref to BWM*	*Introduction of Ballast Water Convention and requests from industry*

PART B - SUBDIVISION AND STABILITY

Regulation	Subject	Amendment	Reason
B 4.1 4.2	General applicability	Footnote excluding specified Cargo ships now in Regulation text Parts B-1 to B-4 apply to PAX ships; application to Cargo ships specified	Footnotes do not have same legal authority as Regulation text
B 4.1	*General applicablity*	*Entire table and footnote on OBO ships deleted*	*All footnotes now included in SOLAS text*

PART B-1 - STABILITY

Regulation	Subject	Amendment	Reason
B-1 5.1	Specifies parameters determined by IE	Displacement LCG, TCG and VCG to be determined by IE	elements of stability' in original text too vague
B-1 5.2	*Cargo ship sister vessels*	*Ref to MSC/Circ.1158 deleted; four new ENs added*	*Clarifies definition of 'sister' ship*
B-1 5.4 and 5	*Alterations*	*New guidance included*	*Clarification of when to incline and when new SIB needed*
B-1 5-1	Information to the Master	Limiting KG_f / GM_T data now to include trim	Expansion and clarification, particularly of Reg. B-1 5-1/4
B-1 5-1.3	Consolidation of Critical KG_f's	New Reg. Full operating range of draught and trim for critical KG_f / GM_T	Intact and Camage Curves sometimes separate with incomplete range
B-1 5-1.3	*Applied trim values to coincide*	*New EN Coinciding trim values applies to intact stability as well*	*Ensures full coverage of intact and damage crit KG_f's for all trims*
B-1 5-1.4	*Development of Limiting GM_T curves*	*Linear interpolation between draughts on limiting GM_T Envelope Curves*	*Clarification on dealing with trim*
B-1 5-1.4	*Development of Limiting GM_T curves*	*New ENs 3 to 5 added covering Envelope Curves, light service and deep T*	*Clarification; dealing with trim in KG_f / GM_T limit curves etc.*
B-1 5-1.5	Alternative to single envelope curve	New Reg. Calculate additional trims with one common GM at each dr.	A trim limit diagram may be more useful than a GM_T Limit Curve
B-1 5-1.5	*Alternative to single envelope curve*	*Expands on new regulation, clarified with diagrams*	*A trim limit diagram may be more useful than a GM_T Limit Curve*
B-1 5-1.6	When KG_f/GM_T Limit Curves not provided	This was old Reg. 5-1.5, reworded eg 'provided' not 'appropriate'	Clarification; now includes non provision of KG limit curves
B-1 5-1.6	*When KG_f/GM_T Limit Curves not provided*	*Cases where Admin approves alternative means of verification*	*Clarification of when this provision may be required.*
B-1 6.2.3	Required subdivision index R	R formula modified to increase R for PAX ships; Ls and LSA removed	Research showed that R was too low; no change to Cargo ship R
B-1 6.2.4	Required subdivision index R; N	Lesser value of N for reduced degree of hazard removed	Too much uncertainty over interpretation of 'impracticable'
B-1 6.2.4	*N relaxed based on Degree of hazard*	*Deleted*	*Explanation no longer needed following deletion of regulation*
B-1 7.1	*Attained Index A*	*New EN's 7 to 10 added*	*Clarification of when new A calculations are needed*
B-1 7.2	Attained Index A	Treatment of trim in calculation of A	Increased clarity
B-1 7.2	*Effect of trim on A*	*Much reduced here but moved to EN B-1 5-1.4-6 in a modified form*	*More rational to included these notes under EN B-1 5-1*

PART B-1 - STABILITY continued

Regulation	Subject	Amendment	Reason
B-1 7.3	Residual *GZ* calculation	Constant displacement removed; free trimming added	Added weight method can be used during IS flooding
B-1 7.5	*A index calculation at curved WL ends*	*New EN3 inserted, with added diagram; old EN3 now EN4*	*Guidance on treating penetration with curved waterlines*
B-1 7.6	*Free surfaces*	*New EN included with cross-reference to EN for R under Reg. 7-2.2*	*How to deal with free surfaces in all stages of flooding*
B-1 7.7.1	*Pipes / valves near bulkhead / deck*	*Now relaxes 'directly adjacent' to 'as close as practicable'*	*Clarification; occurs in several places; only before 01/01/2024*
B-1 7.7.2	*Pipes / valves near bulkhead / deck*	*Now specifies minimum distance of valve from bulkhead with diagram*	*Further clarification for ships after 01/01/2024; requested by IACS*
B-1 7.7.3	*Minor progressive flooding*	*Changes pipe cross-sectional area; new breakpoint 150 metres*	*Different requirement needed for larger ships*
B-1 7-1.1.1.3	*Longitudinal subdivision of zones*	*Length Ls defines the extremities of the zones*	*Clarification*
B-1 7-1.1.1.7	*Pipes / valves near transverse bulkhead*	*Now relaxes 'directly adjacent' to 'as close as practicable'*	*Clarification; occurs in several places; only before 01/01/2024*
B-1 7-1.1.1.8	*Pipes / valves near transverse bulkhead*	*Now specifies minimum distance of valve from bulkhead with diagram*	*Further clarification for ships after 01/01/2024; requested by IACS*
B-1 7-1.1.2.4	*Transverse subdivision*	*Two new ENs with diagrams added (4.1 and 4.2)*	*Clarification of b; treatment of complex waterlines aft*
B-1 7-1.1.2.12	*Pipes / valves near longitudinal bulkhead*	*Now relaxes 'directly adjacent' to 'as close as practicable'*	*Clarification; occurs in several places; only before 01/01/2024*
B-1 7-1.1.2.13	*Pipes / valves near longitudinal bulkhead*	*Now specifies minimum distance of valve from bulkhead with diagram*	*Further clarification for ships after 01/01/2024; requested by IACS*
B-1 7-2.2	Calculation of the factor s_i	$s_{intermediate}$ could now also be applied to cargo ships with Cross-flooding	It was felt that safety standards for Cargo ships should be raised
B-1 7-2.2.3	*Sequential flooding*	*Extensively expanded with new sub-paragaphs and examples / diagrams*	*Clarification of complex calculation procedures needed*
B-1 7-2.2.4	*Cross-flooding*	*Extensively expanded; includes alternative procedures*	*New MSC.362(92) replaces MSC.245(83); clarification*
B-1 7-2.3	Calculation of the factor s_{finali}	Introduces differentiation between RO-PAX and PAX ships; 20/20 factor	Attempts to make some allowance for water-on-deck (WoD) effect
B-1 7-2.3	*RO-RO sfinal, i*	*New EN inserted*	*Explains use of* T_{GZmax} *and* T_{range}
B-1 7-2.4	*Displacement to use*	*Deleted*	*Now in revised SOLAS text*
B-1 7-2.4.1.1	*Definition of breadth*	*Deleted*	*Now in revised SOLAS text*
B-1 7-2.5	Minimising unsymmetrical flooding	of Passenger ships and the freeboard deck of Cargo ships	Clarifies application to Cargo and PAX ships - now applies in many Regs
B-1 7-2.5	*Cross-flooding in Cargo ships*	*Deleted*	*Now in revised SOLAS text for Reg. B-1 7-2.2*
B-1 7-2.5.2.1	Immersion giving $s = 0$	Now applies to Cargo ships only for ships built 01/01/2020 to 01/12/2023	See SDC 6 outcome Paragraph 4.19; conflict with 17.1
B-1 7-2.5.2.3	Immersion giving $s = 0$	New Reg. for PAX ships only built 01/01/2020 to 01/12/2023	See SDC 6 outcome Paragraph 4.19; conflict with 17.1
B-1 7-2.5.2.1 & .3	*Weathertight openings*	*Addition explaining application dates of these Regulations*	*See SDC 6 outcome Paragraph 4.19; conflict with 17.1*
B-1 7-2.5.3.3	Immersion giving $s = 0$	Immersed pipe / duct in damage extent if risk of progressive flooding	Expansion and clarification
B-1 7-2.5.3.4	Immersion giving $s = 0$	New Reg. for PAX ships only built on or after 01/01/2024	See SDC 6 outcome Paragraph 4.19; conflict with 17.1
B-1 7-2.5.5	Watertight hatch covers	Reference to 'small watertight hatch covers' removed	Small' undefined; 'watertight' now applied to all hatch covers
B-1 7-2.5.5	Openings to be kept closed	Delete 'at sea'; replace with 'during navigation in acc with Regulations 22 to 24	For consistency with stated regulations
B-1 7-2.6	*Definition of horizontal WT boundaries*	*Figure illustrating determination of 'v' improved*	*Improved clarity*
B-1 8.1	Fore end damage (PAX ships $N >= 400$)	Must achieve $s_i = 1$ for specified fore end damages, 3 draughts all trims	Clarification of extent of fore end damage and draughts / trims used
B-1 8.1	*Fore end damage (PAX ships* $N >= 400$*)*	*Added EN*	*In case large compartment located aft of collision bulkhead*
B-1 8.2	Si at least 0.9 for side damages $N >= 36$ (PAX)	Specifies draughts and trim for which this s value must be achieved	Clarification of draughts / trims to be used for side damages
B-1 8.3	Extent of damage for compliance with 8.2	Removes references to N and Ls; now 'persons carried N' and L	Consequential upon changes to R formula in Reg. B-1 6.2.3
B-1 8-1.3.2	Provision of onboard computer / shore base	Requires PAX ships built before 01/01/2014 to have computer by 01/01/2025	Improved safety possibly linked to *Costa Concordia*

PART B-2 - SUBDIVISION, WATERTIGHT and WEATHERTIGHT INTEGRITY

Regulation	Subject	Amendment	Reason
B-2 9.1	*Fitting a Double Bottom*	*EN 2 completely re-written; compliance with 9.8 not equivalent to DB*	*Reports that 9.8 damage compliance being used to avoid fitting DB*
B-2 9.2	*DBm should extend to full width of ship*	*New EN 1 explains that not extending to side is 'unusual'; diagram*	*Clarification*
B-2 9.3.1	Bottom of small well min h/2 or 500 mm AB	Additional text to say that if min ht not achieved Reg. 9.8 damage applies	Clarification
B-2 9.3.2	Treatment of other wells (e.g. LO under ME)	Must provide equivalent protection to DB (compliance with 9.8 damage)	Administrations unsure of what constituted 'equivalence'
B-2 9.6	Where no DB fitted PAX or Cargo ship > 80m	Shall withstand bottom damages under 9.8; Cargo ships < 80 to Admin satisfaction	Clarification but also see new EN for Cargo ships < 80 m
B-2 9.7	Unusual bottom arrangements	Differentiates between Cargo ship >80 m and PAX (meet 9.8); <80 m Admin satisf.	Admins unsure of how to treat Cargo ships < 80m (see new EN)
B-2 9.3.2.2, 9.6/.7	*Equivalent safety level; cargo ships<80m*	*Full explanation of how Cargo ships < 80m can achieve equivalence*	*Administrations unsure of what constituted 'equivalence'*
B-2 9.8.2	Assumed extent of damage	Vertical extent AK to be B/20 but within range 0.76 m to 2.0 m	Limits vertical extent to reasonable figure for ships with large *B*
B-2 9.8	*Explanation of 'all service conditions'*	*For B-1 ships use 3 conditions and trims for A; other Cargo ship same as crit KG*	*Administrations unsure of what constituted 'all service conditions'*
B-2 12	*Bow doors*	*Deleted*	*Now incorporated into SOLAS text for Reg. B-2 12.7 and 12.8*
B-2 12.2	s_i not to be <1 at deepest draught	New Reg.; applies for damages forward of collision bulkhead without vertical limit	Extra safety measure for bow damages
B-2 12.6.1	Collion bulkhead penetration	Penetration pipe may be fitted with a butterfly valve; Cargo ships	Consolidates a common practice into SOLAS; ships before 01/01/2024
B-2 12.6.1	*Cargo ship butterfly valves*	*For cargo ships built 01/01/2020 to 31/12/2023 diagrams show butterfly valves*	*Clarification showing acceptable butterfly valve arrangements*
B-2 12.6.2	Collion bulkhead penetration	New Reg. penetration pipe to be fitted with remote valve; all ships	If constructed after 01/01/2024 to have remote controlled fail-safe valve
B-2 12.7 and 12.8	Vertical extension of collision bhd	Both regs now incorporate old EN for Reg. 12 regarding bow ramps	Former EN is now a statutory requirement within SOLAS
B-2 12.10	*After peak bulkhead and RQD for Cargo ships*	*New EN with diagrams showing termination deck for AP bulkhead etc*	*Clarifies common practice on Aft ER bhd and Cargo ship with RQDs.*
B-2 12.11	*Inboard end of stern tube in Cargo ships*	*New EN permitting stern tube to extend through AP/ER WT bulkhead to ER*	*Confirms common practice on Cargo ships*
B-2 13	Watertight bulkheads	The word 'bulkheads' is replaced throughout with 'boundaries'	Now applies to decks and other boundaries below bulkhead deck
B-2 13.2.3	*Use of lead and other heat sensitive matls.*	*New EN; avoidance of progressive flooding in open / closed systems*	*Clarification of testing and approval procedure for pipe penetrations*
B-2 13.3 to 13.13	Power-operated sliding WT doors	Old Reg. 13.6 deleted; remaining Regulations renumbered and revised	Consistency with Parts B and B-1; central operating console for SWTD
B-2 15.9	Gangway, cargo and fuelling ports to be WT	This Reg. now applies only to ships constructed before 01/01/2024	See revised Reg. 15.10 below
B-2 15.10	Gangway, cargo and fuelling ports to be WT	For ships constructed on or after 01/01/2024, old 10.1 and .2 deleted, new R10	Doors to have same degree of WT as shell and open outwards
B-2 15-1.1 to .3	*External openings in cargo ships*	*Application only to Cargo ships subject to Part B-1 damage and other IMO*	*Clarification; will not apply to Cargo ship < 80m for example.*
B-2 16	Testing of watertight closures	Title changed; 'closures' replaces 'WT doors, sidescuttles etc.'	Simplification to make Reg. application more general
B-2 16	*General design standard only*	*New EN states Reg. 16 does not require non-WT hatches to be WT*	*Clarification that introduction of hatches does not over-rule ICLL*
B-2 16.1.1 to .3	Design of doors etc to Admin's satisfaction	Hatches now included as closures; rubbish / ash chutes deleted	Reflects changes in environmental regulations
B-2 16.2	Hatches also to be tested	'hatches' added throughout. New testing requirement for Cargo ships < 80m	Hatches previously omitted; clarification for Cargo ship < 80m
B-2 16.2	*Testing of watertight doors*	*Old EN 1 deleted.*	*Provisions now included in revised Reg. 16.2*
B-2 16-1.2	Vents through the bulkhead deck	Added 'PAX ships', 'watertight area'; removed 'allowable' and 'IS'	Bhd deck not completely WT; 'flooding' more general than 'IS'.
B-2 17.1	WT integrity of PAX ships above bhd deck	This provision now applies only to ships constructed 01/01/2009 to 01/01/2024	Part of attempt to achieve consistency with B-1/7-2.5
B-2 17.1	*WT integrity of PAX ships above bhd deck*	*Old EN for 17.1 replaced with 4 new EN and a diagram.*	*How to deal with WT sliding doors with reduced head/intermittent*
B-2 17.2	WT integrity of PAX ships above bhd deck	New provision applies only to ships constructed on or after 01/01/2024	Part of attempt to implement old requirements in probabilistic terms
B-2 17.2	*WT integrity of PAX ships above bhd deck*	*3 new ENs for new 17.2*	*Inter alia deals with WT doors that are also required to be fireproof*
B-2 17.3	WT integrity of PAX ships above bhd deck	New provision applies only to ships constructed on or after 01/01/2024	Doors above worst IS or final stage for damages contribting to *A*
B-2 17.3	*WT integrity of PAX ships above bhd deck*	*3 new ENs for new 17.3 including a diagram and a table*	*Inter alia deals with WT doors that are also required to be fireproof*

PART B-2 - SUBDIVISION, WATERTIGHT and WEATHERTIGHT INTEGRITY continued

Regulation	Subject	Amendment	Reason
B-2 17.5	Air pipes within a superstructure	Now considered unprotected if no watertight means of closure fitted	Consequential upon switch to probabilistic SOLAS
B-2 17.5	*WT integrity of PAX ships above bhd deck*	*Old EN 17.3 replaced; new EN covers possible progressive flooding*	*Progressive flooding through air pipes to be considered.*
B-2 17-1.1.1	Integrity of the RO-RO deck	Accesses from RO-RO deck below bulkheadd deck - old Reg. re-worded	Improved clarity
B-2 17-1.1.2	Ramps on RO-RO to be closed weathertight	Alarms and open/shut indicators on bridge. WT if horizontal boundary in 7-2.6	Clarifies that ramp openings to be WT if so considered in 7-2.6
B-2 17-11.2	*If ramp assumed to restrict water for A*	*New EN to clarify that non-watertight ramp should comply 7-2.5.3.4*	*s=0 if non-WT ramp opening immersed and counted as buoyant*
B-2 17-1.1.3	Accesses permitted if needed for working	Such accesses to be fitted with bridge alarms and O/S indicators	Slight change of wording from original to include O/S indicators
B-2 17-1.1.1 & .3	*WT doors in bkhds separating RO-RO space*	*Regs 17-1.1.1 and .3 only apply to access from RO-RO deck below*	*Clarifies, with diagram, that horizontal access subject to Reg. 23.3*

PART B-4 - STABILITY MANAGEMENT

Regulation	Subject	Amendment	Reason
B-4 19.2	Open watertight doors in passenger ships	Deleted	Provision now conditional; see MSC.1/Circ.1570
B-4 19-5	DC info to include ref. to onboard computer	New provision for ships constructed on or after 01/01/2024	For alignment with Reg. 8-1.3 re computers or shore-based support
B-4 19-1	Damage control drills	Whole new provisions for passenger ships	Consequences of *Costa Concordia* loss
B-4 20	Loading of passenger ships	'Passenger' deleted	Now applies to all ships
B-4 20.1	Loading of ships	Ship to be upright; loading in accordance with SIB permitted	Initial list not permitted; confirmation of current practice
B-4 21.1 and 4	Operation of watertight doors	'drills' replaced with 'operational tests'; ash and rubbish chutes deleted	More appropriate wording
B-4 21.1	Leaving port	'leaving port' replaced with 'voyage commences'	Port' is too specific; replaced in several Regulations
B-4 22	*Use of 'port' in this regulation*	*'port' includes all berths and sheltered locations for loading/discharge*	*Port' is too specific; replaced in several Regulations*
B-4 22.3	Open watertight doors	Footnote added referring to new guidance in MSC.1/Circ.1564	See also B-4 19.2
B-4 22.3	*Open watertight doors on cargo ships*	*New EN; Guidance not applicable; authorisations up to Admin*	*Relaxation for cargo ships deemed necessary*
B-4 22.5	WT doors dividing cargo spaces	Between deck spaces' changed to 'on 'tween decks'	Thought incorrect in earlier versions of SOLAS
B-4 22.6	Closure of gangway, cargo and fuelling ports	This now only applies to ships constructed before 01/01/2024	See new Regulation 22.7 for ships constructed after 01/01/2024
B-4 22.7	All WT hatches now to be closed	For ships constructed after 01/01/2024. Hatch may be open for limited period	Important to include WT hatches in list of closures at sea
B-4 22.7	*All WT hatches now to be closed*	*New EN applicable to hatches above / below the bulkhead / freeboard deck*	*Clarifies that this applies to any hatch considered WT for damage*
B-4 22.16	Reference to rubbish chutes	Deleted (old 22.17)	Rubbish chutes no longer fitted
B-4 23.6	Essential working of the ship (ro-pax)	Cargo movement during navigation not considered 'essential working'	To limit opening of accesses
B-4 25-1	Water level detectors in multi-hold Cargo ship	New Reg. requiring WL detectors be fitted on Cargo ship built after 01/01/2024	Recommended by USCG following loss of *El Faro*

PART C - REGULATION 35-1 BILGE PUMPING ARRANGEMENTS

Regulation	Subject	Amendment	Reason
C 35-1 3.4	Determination of bilge pump numbers	All flooding conditions' now limited by reference to Reg. B-1 8	Original text gave rise to too many Loading Conditions
C 35-1 3.10	Prevention of flooding from bilge suction	Deepest subdivision load line taken as deepest subdivision draught	Clarification